# ICE FREE

## ELECTRIC VEHICLE TECHNOLOGY FOR BUILDERS AND CONVERTERS

By John Hardy

Cover design by Dave Robinson
Book design by Dave Robinson and John Hardy
Cover photograph by Peter Ohler

Published by Tovey Books
Visit www.tovey-books.co.uk

Printed in various countries

First English Language Edition: February 2012

ISBN: 978-0-9571495-0-2

For my son, Asher, who at the age of 15 came out with a perfectly formed and complete English sentence "Dad, you really should read more Shakespeare"

For Professor John B Goodenough of the University of Texas, Austin, credited with inventing the Lithium Cobalt cell. For good measure, the group that he leads also proposed its successor, the Lithium Iron Phosphate cell.

For the battery makers of China who gave us Lithium ion batteries that a private individual could buy with money and which were big enough to power an EV
（为我们私营散户提供足够驱动 EV 的中国锂离子电池制造商）

*We used to be a source of fuel; we are increasingly becoming a sink. These supplies of foreign liquid fuel are no doubt vital to our industry, but our ever-increasing dependence upon them ought to arouse serious and timely reflection.*

Winston Spencer Churchill  (Parliamentary Debate 24 Apr 1928)


*"The internal combustion engine, one of the greatest technological advancements in history, has an unfortunate downside, namely air pollution so thick that, very soon, sixty-four packs of crayons will include the colour Sky Brown"*

Cuthbert Soup, *A Whole Nother Story*

# Acknowledgements

Many people have contributed to this book, giving their time and their knowledge with great generosity. Some have read sections of the book in draft form and commented on them. Some have shared general information pertinent to the book. Some have agreed to my using their intellectual property. I am grateful to them all: any errors that remain are my fault, not theirs.

I'd like in particular to thank the following:

- Janet Coley and colleagues in the National Grid, for reviewing the section on the impact of EVs on the generation of electrical power
- Professor Derek Fray FRS, FREng for sharing some insights into developments in Lithium Ion electrode materials, and reading and commenting on the text.
- Huw Hampson-Jones, Daniel Ugbo and Scott Lilley of Oxis Energy for taking time out of their busy work schedule to improve my understanding of advanced batteries.
- Dr Austin Hughes, formerly Department of Electrical and Electronic Engineering, University of Leeds, UK for reading and commenting on the motor and controller  chapters
- Professor Willett Kempton of University of Delaware for reviewing the section on Vehicle-to-Grid.
- Clive Lingard for reading and commenting on sections of the text
- Chris Rogers of Network Operations at the National Grid for supplying the data underlying Figure 9-10 and 9-11.
- Andrew Vallance of Protean Electric for contributing most of the section on wheel motors and for input on other motor-related areas.
- Dr Alan Ward (former Chairman of the Battery Vehicle Society) for wading through the entire book in draft form and suggesting a number of improvements.

I owe a great debt to the scores of kindly individuals and a number of commercial organisations[1] who have supplied photographs or given me permission to use photographs that they have published elsewhere.

I have attributed these photos in the text. I believe I have done this accurately but if you spot an error I would be most grateful if you would let me know so that I can make a correction in future editions. If I have wrongly attributed your photo to someone else I would in addition crave your forgiveness.

I would also like to acknowledge a set of creative people who are often overlooked: the designers of the fonts (typefaces) used in this book. The shape of each individual, letter, numeral, punctuation mark etc in this book was conceived and honed by a designer. The text that you are reading now is Liberation Serif. The bulk of the text is set in Liberation Sans. Both these fonts were designed by Steven R. Matteson. **FRESHMAN** was used for the title page. This font was designed by William Boyd.

Finally I'd like to salute Jack Rickard. Jack is a wealthy American who conceals a brilliant mind and a good heart behind a bumbling and mildly abrasive down-home demeanour. He spends his time and money converting cars to electric drive and producing a free-to-view weekly web-only TV show (http://www.evtv.me/) which is a mixture of electric vehicle news, a documentary of his conversions, and primary research on batteries. Some of the battery tests that he has done (noting his data on the backs of old envelopes) have come up with results I've seen nowhere else.

John Hardy
January 2012

---

[1] Figure 7-10 is provided by Tecumseh Products Company, contact Tony Carstensen, Masterflux Business Unit Director, Tecumseh Products Company, 1136 Oak Valley Drive, Ann Arbor, MI 48108; (734) 584-9475, http://www.masteflux.com, tc@masterflux.com

"By giving permission for the use of product photographs, Tecumseh Products Company makes no representations or warranties whatsoever regarding use of Masterflux® compressor products for the purposes discussed in this work."

# Contents

# Table of Figures

# 1: Introducing Electric Vehicles

Electric vehicles (EVs) have a lot going for them: little noise apart from the rush of air and the whine of tyres. No urban air pollution. Little dependence on dwindling global oil supplies. Negligible fuel costs. No need for a network of fuel stations. Low speed acceleration that humbles the mightiest supercar. And (if it matters to you) lower carbon emissions.

Of course EVs have their problems too – primarily poor electric-only range, expensive batteries and time-consuming recharge. These problems have meant that electric vehicles were only just reaching mass-market early adoption in 2012, but the price of EVs from mainstream manufacturers is high, availability restricted and the choice limited; so if you want one, you may want to consider building or converting a vehicle yourself. This book will give you the knowledge you need to decide how to go about just such a build or conversion.

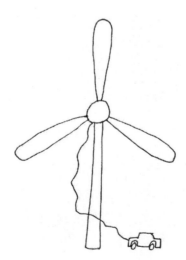

*Figure 1-1 EVs are not fussy about where the electricity comes from*

But first what do we mean by an electric vehicle (EV)? A pure EV is simple enough – a road vehicle propelled by an electric motor rather than an Internal Combustion Engine (I.C.E. - i.e. a conventional petrol or diesel engine). It is usually assumed that the electric power to run the motor comes from onboard batteries (rather than fuel cells, or overhead wires like an old-style trolley bus). Such electric vehicles are sometimes referred to as "BEVs" (Battery Electric Vehicles").

Then there are Hybrid Electric Vehicles (HEVs). Hybrids use two sorts of power – an IC engine plus an electric motor in one of four basic configurations. HEVs are discussed in more detail later.

# Strengths and weaknesses

The best part of an electric vehicle is the motor. A cheap electric motor is superior to the mostly highly developed conventional internal combustion (IC) engine in almost every respect: for similar output, electric motors are smaller, lighter, have far fewer moving parts, are much quieter and need far less maintenance (there are some industrial electric motors that have run for over 100 years – for example the Middlesbrough's Transporter Bridge was built in 1910 and the electric drive system was replaced in 2011).

Electric motors don't need an ignition system, fuel system or exhaust system. They never need an oil change. Old ones don't drip on your drive. They are very efficient so don't get nearly so hot. Electric motors have several more subtle advantages. Piston engines have a narrow torque band: if you are driving a hot hatch and want maximum acceleration from a standing start you need to anticipate the need for power, wind up the rpm and slip the clutch off the line. The typical EV electric motor on the other hand is set up to provide maximum torque at or near zero rpm, so a maximum-effort standing start is simply a matter of pushing the accelerator to the floor. This is part of what makes a good EV fun: you are never caught in the wrong gear, because the torque curve looks like Table Mountain rather than Mount Fuji.

Some EVs don't even have a gearbox at all. None need automatic transmissions, flappy paddle gearboxes, lightweight flywheels and all the other paraphernalia invented to get round the limitations of IC engines.

2

Electric motors tend to be high-torque devices. For example, an 11" (28cm) diameter series wound electric motor is a monster that is quite capable of snapping a driveshaft if it is fed enough amps (Figure 1-2).

That kind of motor doesn't need help from a puny little four-banger in accelerating you away from the lights. Moreover it manages to combine neck-snapping acceleration off the line with good economy in cruise. A big powerful IC engine is hopelessly inefficient around town. A big powerful electric motor isn't.

There is however a moment when the puny little four banger can cough modestly and take a step forward. That's when somebody mentions that there are five adults and their luggage to be taken from London to Glasgow. The 11" motor (or even a 7" motor) would have no trouble with such a trip. The issue is batteries. No one has yet marketed a battery that is anywhere close to doing the job. Compared with a tank of fuel, even the best batteries store a pitiful amount of energy and (in practice) take a long time to charge. The all-electric Tesla Roadster is said to have a range of 250

*Figure 1-2 Warp 11 motor - capable of generating huge torque from zero rpm with the right controller. Photo courtesy of Netgain Motors (http://www.go-ev.com)*

miles on a charge – that is phenomenal by EV standards (20 – 90 miles is more common) – but the Tesla has only two seats and a set of batteries that cost as much as many complete cars. When it comes to load carrying over long distances, the best EV in the world is totally outclassed by an early VW Beetle or a Fiat Panda.

So is that it then? Are EVs are so seriously, er, range-challenged as to be useless? Well, not quite. The statisticians tell us that for most people most of the time driving distances are short. How short? Less than 40 miles a day. A 40 miles range is right in the sweet spot for an EV. If you commute 40 miles a day and can persuade your employer to put a charging post in the car park at the office, you only need a 20 – 30 mile range.

So whether EVs are for you depends on your needs. If you are an Account Manager covering a large territory, then an EV wouldn't work for you. If on the other hand most of your driving is moderate-distance commutes and shopping trips, then an EV may be well-suited to your needs.

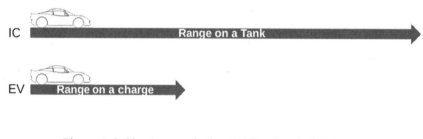

*Figure 1-3 Short range is the Achilles heel of EVs*

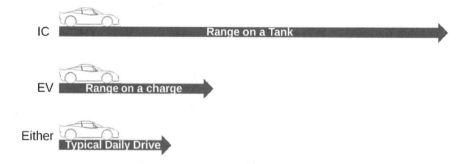

*Figure 1-4 Most people don't need long range most of the time*

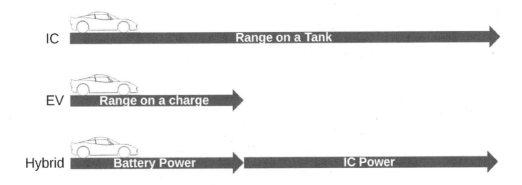

*Figure 1-5 Plug in Hybrids combine the best of both worlds (at some cost)*

So how do you cope with the occasional long journey? If you are a two-car household, one EV and one conventional car may be a good mix, with the choice of who gets which car decided over the breakfast table. If you only have one car, but you go and visit granny on the other side of the country a few times a year, it may actually work out cheaper to have an EV as your daily drive and rent a conventional car for the odd occasions that you need it. Check it out. A forty-mile commute five days a week is 200 miles a week – maybe 6 gallons of petrol a week, or 24 gallons a month. At the time of writing, petrol is about £6.00 a gallon in the UK, so 24 gallons is £120. Weekend hire of a full size car might be around £50. Even when you factor in the cost of electricity (perhaps 200 kW-hours at 10p/kW hour =

£20[1]) it still looks feasible. If you are in the US the numbers will be different, but the conclusions much the same.

Another approach is the leisurely one. A long distance drive in an EV is perfectly possible if you are prepared to take long breaks. If for example you have a 150 mile journey to do and an EV with a range of 100 miles, then you may find that you could phone ahead and find a pub that could offer you an electrical outlet whilst you have lunch. EV Network UK (http://www.ev-network.org.uk/) maintains a list of public charging points and a members-only list of fellow-EVers willing to offer charging facilities. Figure 1-5 illustrated a third option

---

[1] "kW" is used throughout this book as an abbreviation for "kilowatt" (1000 Watts)

– convert your car to a plug-in hybrid rather than a pure EV. Most of the time, leave the IC engine switched off and use the car like an ordinary EV. When you are going on a longer trip start with charged batteries but use both sources of power to stretch the range. Hybrids are not all sweetness and light however. They need a complete electrical drivetrain, plus all or most of the elements of an IC drivetrain. Hybrids are therefore more expensive than either an equivalent IC-engined car or an equivalent EV. Furthermore (all other things being equal) they are heavier and have less room for people and luggage.

# The shape of this book

This book is intended to be a foundation. It introduces you to the background knowledge and principles that you need in order to build an electric car, or (more likely) to convert a car originally powered by a petrol or diesel IC engine to electric drive. It is not a detailed "how to" book, neither is it a substitute for researching what is out there: the field is moving too quickly for that. What it will give you though is enduring principles that can be applied to products that will appear in the coming years. We'll minimise the maths, but to get the best out of the book it will help to understand volts, amps, watts, resistance and the concepts of energy, torque and power. If you are a little hazy on some of these, never fear: the next chapter is a refresher which you can skip or skim if you are comfortable with the theory.

Brushed Permanent Magnet DC

Series wound and Separately-excited DC

"Brushless DC"

AC Induction

*Figure 1-6 Various Motor Types, Photos courtesy of the following - left to right: Bruce Roberts, Netgain Motors, Guangdong M&C Electric Power Co., MES-DEA.*

Thereafter, we'll start with *motors*. There are several types of motor that could be actively considered for EV conversions (see Figure 1-6). By the time you have read the chapter on motors you should understand the principles behind all these motor types.

Low power DC controller

High power DC Controller

Controller for AC motors

*Figure 1-7 Typical Controllers. Photos courtesy of the following - left to right: Curtis Instruments, Kurt Clayton, BRUSA Elektronik AG*

Next we'll look at *controllers* (Figure 1-7). Controllers are to EVs what carburettors are to petrol engines. In an EV, the throttle pedal is connected to the controller. The type of controller is governed by the type of motor, but a poor choice of controller can either waste a lot of money or turn your car into a gutless wonder.

AGM Lead Acid

Flooded Lead Acid

Prismatic Lithium Ion

Nickel Metal Hydride

Lithium Sulphur

Lithium Ion Polymer

*Figure 1-8 Battery Types. Photos courtesy of the following (top row, left to right) OPTIMA Batteries, Trojan Battery, Greg Sievert (bottom row) Randy Hsieh, Andrej Pecjak*

After that we will move on to *batteries* (Figure 1-8). Batteries are probably the largest, heaviest and most expensive element in a conversion.

The section on batteries is in two parts. The first will focus on the principles of operation of the batteries themselves. The second will look at the thorny issue of managing long strings of batteries in an EV, including how to charge them.

When we have concluded the section on batteries we'll move on to the other things that you need to consider if you are going to have an EV that you can live with summer and winter. These include heating, air conditioning, power steering and power brakes.

Next we will look briefly at some of the other technologies that have been discussed as alternatives to IC engines for powering road vehicles. These include ultra-capacitors and even hyper-flywheels and compressed air. We'll also look at fuel cells and outline why, in the short term anyway, this technology is a monstrous red herring. In this section, we will look in more detail at the four classes of hybrid, ranging from machines that are little more than an IC engined car with an electrically-augmented transmission, to EVs with a range-extender.

Widespread adoption of EVs could be expected to have positive benefits in unexpected ways. We'll discuss the evidence for the likely impact on such diverse areas as childhood stress, and the severity of asthma attacks in urban areas. The $CO_2$ emissions associated with EVs will be covered, but there will be no preaching about climate change. We'll also talk about the risks. Oil and the automotive sector constitute a large chunk of world economic activity. Changing it radically is risky. Also a widespread move to EVs might just exchange one materials shortage (oil) for another (battery or motor raw materials).

NB where prices or price ranges are quoted in this book, they are approximate, illustrative and at the time of writing.

# 2: Refresher course

I promised to minimise the maths, but if the rest of this book is to make sense, you need to be equipped with a few concepts, so we will start with a refresher on these. If you know all about this kind of thing then feel free to skip the chapter. The topics we will cover include:

- Volts amps and ohms
- Series and Parallel circuits
- AC and DC power
- Back EMF
- Electricity and Magnetism
- Force, Torque, Energy and Power

## Volts, Amps and Ohms

One of the most helpful ways of visualising electricity is to think of it like the flow of water in a pipe. Water flow has two principle characteristics:

- *Pressure* (measured in bar or pounds per square inch)

- *Volume* or *mass flow* (in litres per hour or gallons per minute)

To take an example, a pressure washer is a high pressure/low volume device. It pushes out relatively little water but does so with great force. On the other hand a culvert or drain pipe buried under a road may pass a vast amount of water but at relatively low pressure. A fire hose is somewhere in between: it produces quite a lot of water at a fairly high pressure.

*Voltage* in an electrical circuit is a bit like water pressure: it is what makes the current flow. Electrical *current* on the other hand is like the amount of water flowing in the pipe. An automotive ignition system is a bit like that pressure washer: it delivers enormously high voltages but the current (the actual amount of electricity flowing) is tiny. An electric arc welder on the other hand is like the culvert: an enormous

9

*Figure 2-1 High pressure, low volume (top); low pressure, high volume (bottom). Photos courtesy of Auto Lavish (top left), Rob Clare (top right), Jen Cox (bottom left), Official US Navy images under the Creative Commons Attribution 2.0 Generic licence (bottom right)*

current at relatively low voltage (Figure 2-1)

So an electrical voltage corresponds to water pressure, and the electrical current measured in amperes (or "amps" for short) corresponds to the volume of water flowing. So how do the two relate? Once again,

there is a useful analogy with water in a system of pipes: the more restrictive the pipes, the lower the flow. The higher the pressure, the greater the flow.

It is the same with electricity. The electrical equivalent of a restriction in a pipe is

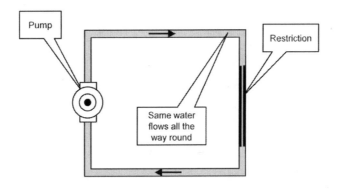

*Figure 2-2 Water "circuit"*

*resistance*. For a given resistance, doubling the voltage doubles the current, and halving the voltage halves the current.

The other thing about electric current is that it normally needs a complete loop or circuit, and the current is the same all the way round the circuit. Once again, the analogy with water flowing in a pipe helps visualise it (Figure 2-2). Unlike a water pipe however,

a break in an electrical circuit does not result in electricity spilling out onto the ground!

Here for comparison (Figure 2-3) is a simple electrical circuit with a battery in the left hand leg and a fixed resistance in the right hand leg. As with the water, the same current flows at all points around the circuit. The concept of electrical resistance is an

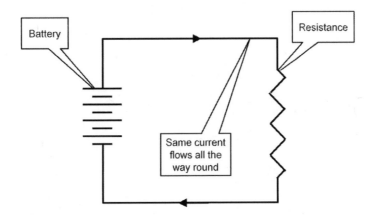

*Figure 2-3 Simple electrical circuit*

important one. Volts and amps can vary, but resistance is usually a fixed characteristic of an electrical device (although it may change with temperature). A motor field winding has a resistance, an electric light bulb has a resistance, even the wires that you use to connect up electrical devices will have a (small) resistance.

Resistance is measured in Ohms (Symbol Ω) and is simply the voltage drop across the device divided the current flowing as a result or

$$R = V \div I \; or \; V = I \times R$$

...where R is the resistance in ohms, V is the volts applied and I is the resulting current in amps: so for example if you have a 5 Ω resistor and put it across a 10 volt battery, you would expect a 2 Amp current (10 volts ÷ 5 Ω) to flow (Figure 2-4).

Alternatively if you measured the current and found that it was 5 amps, you could deduce that the battery provided 25 volts (5 Amps x 5 Ω) - Figure 2-5. Devices intended for use with mains electricity (240 Volt in the UK) have a much higher resistance than devices intended to work off (say) a 12 volt car battery. To take an example, a typical old style incandescent light bulb designed to work from UK household current will have a resistance when hot of around 1000Ω. A typical 12v volt car headlamp bulb on the other hand has a resistance of around 2.4 Ω in use. If you put mains voltage across the car headlamp bulb, the current would be 240/2.4 i.e. 100 amps – for a very short time before it burned out.

These terms are at the heart of electrics and electronics, and in truth they are not difficult concepts.

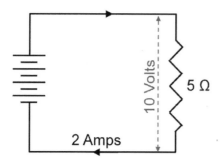

*Figure 2-4 Ten volt battery, five ohm resistance = two amps*

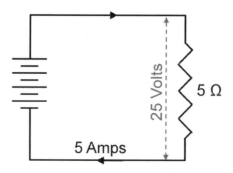

*Figure 2-5 Five ohm resistor, five amps therefore twenty five volts*

# Series and Parallel

If several electrical components are connected so that the current splits and a proportion of the current goes down each one independently, they are said to be "in parallel" (Figure 2-6). So for example the booths at the end of a toll road are the equivalent of parallel electrical components. On the other hand when the components are connected in a chain so that the same current is passing through them, like checkpoints on the road into a city in a war zone, they are said to be "in series" (Figure 2-7).

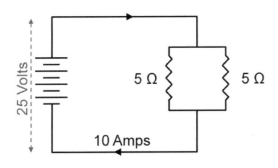

Figure 2-6 Parallel circuit (currents additive)

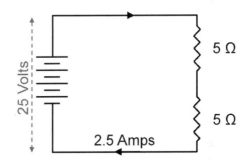

Figure 2-7 Series circuit (voltage drops additive)

# AC and DC

AC stands for "Alternating Current" and DC is "Direct Current". As the name implies, in an AC electrical system the voltage (and therefore normally the current flow) isn't just one way: it stops and reverses direction, typically many times a second. If you think of an electric current being like the flow of water in a river, then AC is like the back-and-forth tidal movement you get near the mouth of the river, only immensely speeded up (Figure 2-8 shows a graph of voltage against time for a typical AC power source). DC ("Direct current") on the other hand is a one-way flow, like water in a hosepipe or a mountain stream.

For reasons that are difficult to explain in a soundbite, AC machines tend to be simpler than their DC equivalents. Critically, you can change the voltage of an AC supply using a cheap and simple transformer with no electronics and no moving parts. For this reason, if for no other, mains electricity is AC throughout the world. In the UK, it operates at 50 cycles per second, in the US at 60.

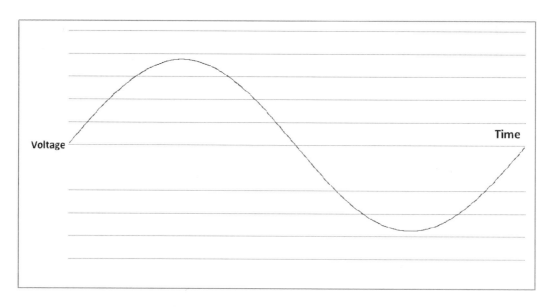

*Figure 2-8 AC voltage varying with time*

Unfortunately, no one has ever produced an AC battery. Batteries are always DC. It is possible however to produce AC from a DC source using a device called an "inverter" (Figure 2-9).

Such are the efficiency advantages of AC that it is sometimes worth going to the trouble of turning your DC into AC before you use it.

*Figure 2-9 - An inverter: takes in DC power from a 12 volt source and outputs mains-voltage AC.*
*Photo courtesy of Black and Decker (http://www.blackanddecker.com/)*

# Electricity and Magnetism

There are just four things that you need to know about the interaction of electricity and magnetism to understand how electric motors work:

1. Moving a wire in a magnetic field (or moving a magnetic field past a wire) creates (or "induces") a voltage in the wire

2. A current flowing along a wire creates a magnetic field around the wire.

3. If you give a magnetic field a choice of going through air or soft iron, it will take the iron every time. You can therefore use iron to shape and focus a magnetic field.

4. A current flowing in a wire in a magnetic field creates a force on the wire

Here are four experiments that you can try out in order to illustrate these principles. You can try them for real if you wish or just do the experiments in your head, or look on YouTube, or on the website for this book.

# Experiment 1 – Inducing a voltage

Find yourself an ordinary magnet (the photo shows a powerful ring magnet) and a piece of wire. Beg or borrow a sensitive electric meter. Connect the wire to the meter. Move the wire around near the magnet.

You should see the meter kick (Figure 2-10). This illustrates that moving a wire in an electric field creates a voltage. This is the basis of all electrical generators. It also happens in motors as well, as we'll see later.

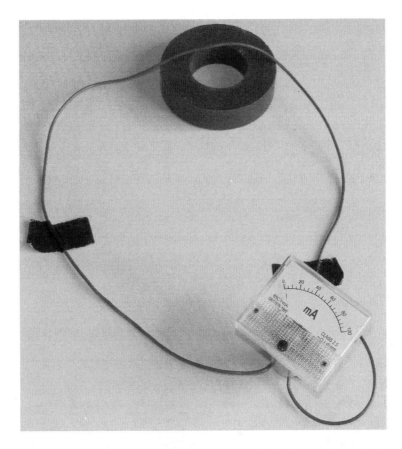

*Figure 2-10 Moving a magnetic field past a wire induces a voltage in the wire*

# Experiment 2 – Magnetic field round a wire

Acquire a coil of fine wire. Connect a battery and a switch in series with the coil. Put a compass near the coil of wire. Close the switch and watch the compass react to the magnetic field created by the current flowing through the coil (Figure 2-11). Note that the magnetic effect probably won't be very strong: probably not enough to lift something like a paperclip.

This is a very simple electromagnet. All electric motors contain a magnet or magnets of some sort. They may be permanent magnets, or they may be electromagnets.

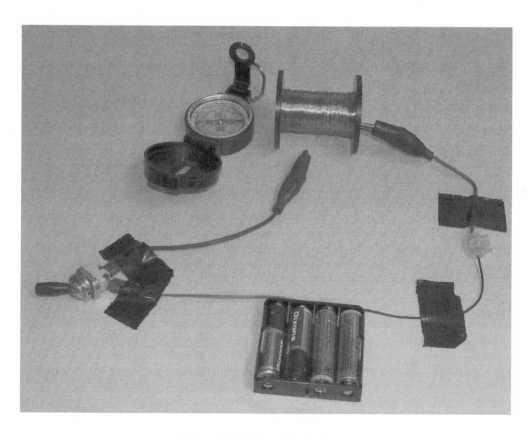

*Figure 2-11 A crude electro-magnet*

# Experiment 3 – Add some iron

Repeat experiment 2, but this time place a big nail or any other solid iron or ordinary steel (not stainless) bar inside the coil (NB steel is just iron with a bit of carbon and sometimes other metals in it to make it stronger). I used a piece of threaded rod

What you should see this time when you connect the battery is that the magnetic field is much stronger and focussed particularly at the ends of the bar. If you

have sufficient number of turns of wire and enough current flowing, you should be able to use the metal bar like a permanent bar magnet to pick up other iron and steel objects.

This experiment illustrates why electric motors are so heavy for their (usually small) size: they have lots of iron in them. Electrical transformers are usually heavier than expected too, for much the same reason.

*Figure 2-12 Electromagnet with Iron Core*

# Experiment 4 – the electrical kick

For this experiment drape a piece of wire loosely across the magnet you used in experiment 1 and connect it to the battery and switch that you used in experiments 2 and 3. This time when you switch on, the wire running past the magnet should twitch (Figure 2-13). The interaction between the electric current and the magnetic field applies a force to the wire. This is the principle of most electric motors as we will see later.

*Figure 2-13 A current flowing in a wire in a magnetic field generates a force*

# "Back EMF"

In experiment 1 ("Inducing a voltage") we demonstrated that moving a wire in a magnetic field creates a voltage in that wire. As we shall see in more detail later, a DC motor consists of a number of wires wound around a rotor, which rotates in a magnetic field. Regardless of what else is going on, this creates a voltage in those wires. In fact if you take a DC electric motor and rotate the shaft using an external power source, you have got yourself a generator: instead of the current flow through the motor producing power, power is absorbed as a voltage is created and a current flows.

You can actually think of this "generator" effect happening all the time in electric motors, even when the electric motor is working normally: in other words the rotation of the motor creates a voltage pushing back against the externally applied voltage from the battery. In Figure 2-14 for example 25 volts is being applied to a motor, but the rotation of the motor is inducing 20 volts in the opposite direction. Net, there are just 5 volts (25 minus 20) available to drive current around the circuit.

The faster the motor spins, the higher this reverse voltage will be. Assuming that the motor doesn't exceed its mechanical limits and burst first, the maximum speed of a motor will be reached when the generated reverse voltage (or "back EMF" - Electro Motive Force) reaches the same value as the applied external voltage (minus a little bit for windage and bearing drag). So all other things being equal, we would expect EVs with the highest battery pack voltage to have the highest top speeds.

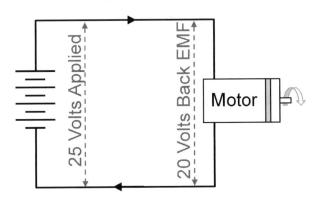

*Figure 2-14 Back EMF builds up to oppose the applied voltage from the battery*

# Force, Torque, Energy and Power

These four terms (Force, Torque, Energy and Power) tend to be used vaguely and often interchangeably yet they have very precise meanings which it helps to understand if you are interested in the dynamics of vehicles. In particular if you are to have any hope at all of estimating the performance potential of a planned EV you need to understand them.

Motoring magazines and armchair racing drivers tend to compare the performance potential of cars using engine power output: 20 Horsepower – bad; 200 HP – good. Road tests often discuss maximum torque. On the road, torque is actually what makes a car feel lively. If you are familiar with these terms I suggest you skip to the next chapter. If not, you might want to read on....

## Force

This is the easy one as most people have a feel for it. It is a push, a shove, an influence which makes a free mass move or a flexible structure change shape. Anything in earth's gravitational field experiences a force, so if you want to hold something still, you have to apply an upward force equal to its weight to cancel out the effect of gravity.

In the metric system, force is measured in *Newtons*. Appropriately, if you hold an apple in the palm of your hand you are applying an upward force to it of roughly a Newton (depending on the weight of the apple).

You can also measure force in *pounds* or *kilograms* (yet another unit of force you might encounter is the *poundal*). Unsurprisingly, a kilogram is the upward force you would have to exert if someone replaced the apple resting in your hand with a 1 kg bag of sugar. It is just less than 10 Newtons.

## Torque

Torque is a measure of how hard something is twisting. A heavy man jumping up and down on the end of a metre-long spanner can produce a lot of torque (probably more than a big V8). Torque is the rotary equivalent of a linear force: in fact it is measured as force x distance (Newton-metres or foot-pounds). So a 5 Newton force on the end of a 10 metre bar creates a torque of 5 x 10 = 50 Newton-metres. Exactly the same torque would be produced by a force of 50 Newtons at the end of a one metre bar (or a 25 Newton force on a 2 metre bar or a one Newton force on a 50 metre bar – or whatever).

Note that torque says nothing about speed of rotation. If you are winding a bucket of water up from an old fashioned-well the torque that you are applying is almost the same regardless of how fast you are raising the bucket (assuming that the bucket is moving at constant speed).

For most road driving, torque is more important than power. For a given gear ratio the shove in your back when you floor the throttle relates to the torque output of the engine. In the real world, for cars of similar weight, the maximum torque of the engine has (rather surprisingly) been found to be a better predictor of 0-60 mph acceleration time than maximum power. Electric motors are generally good at torque, so EVs often have surprisingly good acceleration.

## Power

If you are winding up that bucket of water (Figure 2-15), you would intuitively expect that the faster you pull the bucket up, the greater the physical effort. As the physicist would put it, raising the bucket fast needs more power, even though the torque doesn't change.

Power is torque x rotational speed or (in linear terms) force x linear speed. You come across the same kind of effect when you are push starting a car. Keeping the car rolling at a slow walking pace is no problem, but if you have to get it moving at a brisk trot to bump-start it, you get out of breath quickly. The force that you are applying hasn't changed but keeping up that force at a run demands more power than doing it at a walk.

Calculating power in the metric system is easy: a force of 1 Newton at 1 metre per second is 1 Watt. So a 10 kW (10 kilowatt or 10,000 Watt) power source could in theory apply 1000 Newtons at 10 metres per second or 400 Newtons at 25 metres/second, or whatever (Figure 2-16).

*Figure 2-15 High power expenditure for a short time*

Maximum power relates to top speed, so is very important in racing (less important on the road where cars seldom get anywhere near their maximum speed). As the speed rises, aerodynamic drag and rolling resistance are pushing against the engine. With the right gearing, maximum speed is reached when the forces trying to slow the car down multiplied by the speed of the car equal the maximum available power. At this point something quite remarkable emerges. There is a fearful symmetry in the universe; because not only is 1 watt = 1 Newton at 1 metre per second. 1 watt is also equal to a 1 volt x 1 amp. And it isn't that they just work out to be roughly similar. They are the same thing. So for example a perfectly efficient motor drawing 50 Amps at 200 volts would produce 50 x 200 = 10,000 watts. With the right gearbox and final drive it could provide a force of 400 Newtons to a car travelling at 25 metres per second as in the bottom picture in Figure 2-16.

In other words the world of moving bodies on the one hand and the world of electricity and magnetism are connected. The fact that they are exactly 1 to 1 is partly down to the cunning of the metric system, but only partly. It is enough to fill you with awe that two such different things should be so tightly bound together. It also makes the job of predicting the performance of EVs a whole lot easier. Furthermore the concept

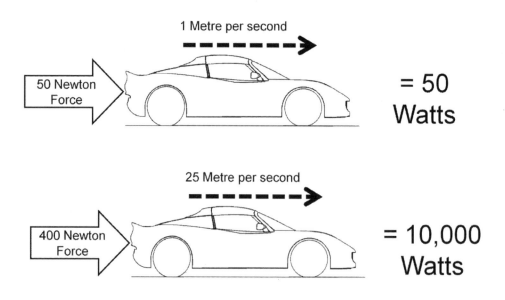

*Figure 2-16 Force, speed and power*

of efficiency of a drivetrain becomes meaningful. Because we measure power *in* to the motor as volts x amps and mechanical power *out* of the transmission as force x speed, we can divide one by the other to get efficiency.

## Energy (or work done)

Energy reflects the application of a given amount of power for a given length of time (power x time). So going back to the bucket of water in the well, the total energy that you expend winding the bucket all the way up from the bottom to the top is constant regardless of whether you wind it up quickly or slowly. If you wind the bucket up very fast you need lots of power but for a short time. Conversely, if you wind it up more slowly, you will need less power, but you will need it for longer. Either way the total energy expended is the same (ignoring friction etc).

In the metric system, energy is measured in *Joules* which is 1 watt for 1 second (i.e. 1 Newton being pushed along at 1 metre per second for a period of one second, or 1 Newton being pushed a metre). In electrical terms it is 1 volt x 1 amp for 1 second. That isn't much energy, so electrical energy is usually expressed in kW-hours. 1 kW-hr is 1000 watts for 3,600 seconds or 3.6 million Joules.

The distinction between energy and power will be very important when we start discussing batteries. But for now we'll take some of the electrical principles we talked about here and apply them to the kinds of motors used in EVs and HEVs. Motors are the subject of the next chapter.

# 3: Motors

Electric motors are at the heart of EVs. It is the electric motor (or motors) which turn electrical power from the batteries into mechanical power to turn the wheels.

Motors are divided into two broad classes – AC and DC machines, but if you understand one class, you are well on the way to understanding the other.

## Electric Motor principles

Electric motors are much simpler than their IC counterparts. Here is a list of the moving parts in a typical 4 cylinder IC engine:

- Crankshaft
- Connecting Rods (x 4)
- Pistons (x 4)
- Gudgeon Pins (x 4)
- Camshaft
- Cam followers (x 8)
- Pushrods (x 8)
- Rocker arms (x 8)
- Valves (x 8)
- Distributor shaft
- Oil pump

We could also mention throttles, chokes, turbocharger turbines, wastegates, cam drive belts etc; but even this basic list is almost 50 moving parts. For a V8 it is close to 100.

Most electric motors have by contrast just two main parts of which one is fixed.

- a stator
- a rotor

Regardless of type, Electric motors rely for their operation on the interplay of electricity

*Figure 3-1 An AC electric motor with the stator built into the case and the rotor on the shaft. Photo courtesy of Stan Zurek*

and magnetism. Stators are usually magnets. They are often a form of electromagnet (electrical wire wound around an iron core like the wire-wound nail in the third experiment described earlier); or they may be permanent magnets. So electric motors are very simple devices. Some are low maintenance. Others are completely maintenance-free (how long has the motor in your refrigerator been running without attention?).

One peculiarity of electric motors is that the maximum torque that they can produce tends to be related to the volume of the motor, regardless of type: so a long thin motor will have similar maximum torque to a short fat one if the volumes are the same; and (in perfectly designed and well-cooled motors) the peak torque will be much the same regardless of motor type.

Electric motors are also small when compared to an IC engine of similar output. The Netgain Impulse 9 (Figure 3-2) has, unsurprisingly, a diameter of about 9" or roughly 22cm. This is a fairly powerful motor, quite capable of pushing a large car or small van along fast enough to keep up with the traffic. In a smaller car it is quite capable of seriously embarrassing IC engined cars at the traffic lights. 9" is only about the diameter of a medium sized cake tin or a small frying pan. Electric motors are usually heavy though. The Impulse 9 weighs about 60 kg. The Warp 11 illustrated in Chapter 1 weighs over 100 kg.

*Figure 3-2 The Impulse 9 Motor. Photo courtesy of Netgain Motors (http://www.go-ev.com/)*

# Types of Motor

There are many different types of electric motor but there are probably five types available for purchase today that are potential candidates for propelling an EV:

- Brushed DC Permanent Magnet
- Brushed DC Series wound
- Brushed DC Separately Excited
- AC Induction
- "Brushless DC" (really a form of AC motor despite its name)

They all have their advantages and disadvantages. The AC motors are simpler devices than the DC machines and are generally more efficient, but they need complex and expensive controllers (which, among other things) converts the DC output of the battery into the AC needed by the motor.

Among practical motors available today the series wound DC motors tend to be king of the hill when it comes to low speed acceleration. Electric dragsters (there are such things, honest) are often driven by Series wound motors. On the other hand brushed permanent magnet motors tend to be small, light and (relatively) cheap. Practical AC systems currently available for EVs are better at regeneration than comparable DC systems; there is no theoretical reason for this, but it appears to be so. "Regeneration" or "Regen" is slowing the car down by using the motor as a generator instead of using ordinary brakes. This puts the energy from the slowing car back into the batteries, rather than dissipating it as heat from the brakes, and thus reduces wear and tear on the brakes. In theory it also extends the range of the EV, albeit not by very much in practice under normal conditions. We discuss regen further in the section on Controllers.

We'll look at the principle of each motor type starting with the simplest: the brushed DC permanent magnet motor.

# Brushed DC Permanent Magnet motor

If you ever played with small battery-operated electric motors as a child, it is probable that the motors you were using were permanent magnet brushed DC motors. They use the principle that was illustrated in Experiment 4: a current passing through a wire in a magnetic field experiences a force. Imagine a rectangle made of wire with a pivot down the middle and a current flowing in it. The wire on one side would experience an upward force and the wire on the other a downward force (Figure 3-3). This would turn the whole rectangle of wire on its axis.

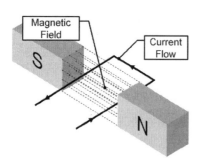

*Figure 3-3 Force on a wire in a magnetic field*

That's all very well, but such a motor would have a useful life of a quarter of a turn: when the rectangle of wire was lying across the magnetic field there would be no torque on it. After half a turn, the torque would be in the opposite direction.

To turn this into a usable motor, you need a commutator and brushes. A commutator is a series of conducting metal strips laid length-wise around the outside of cylinder on the axis of the motor. Each of the commutator segments is insulated from its neighbours and connected to the commutator segment on the opposite side by a wire or bundle of wires. Figure 3-4 illustrates the principle (just one winding is shown for clarity). In a real motor there would be dozens.

A brush is a sliding electrical contact. Brushes are typically made of graphite and are mounted in a housing attached to the motor frame, and are spring loaded to keep them in contact with the commutator

The two brushes feed electrical power to each pair of commutator segments in turn. If you want to see an animation, try searching for "DC Electric Motor" in YouTube: you will probably find several.

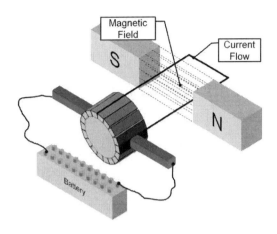

*Figure 3-4 Principle of the Permanent Magnet DC Motor*

The brushes and the commutator are an elegant idea which makes a simple DC motor feasible; but they are also a weak point. Brushes wear out and in a big motor they may be asked to carry a very high current (2000 Amps is not unheard of with series wound motors in EVs). Under these conditions you can get serious arcing which can damage the commutator and/or brushes. Brushes also limit the maximum operating speed of the motor. The best-known EV permanent magnet motors are those made by Agni motors to a design by Cedric Lynch[2] - Figure 3-5.

These motors are almost unbelievably small and light; they are the size and shape of a large tin of biscuits. The largest in the range has a peak power output around 30 kW (38 HP) and weighs just 11 kg. Two of them would make a small hatchback quite frisky.

*Figure 3-5 Agni Brushed Permanent Magnet Motor bolted to a rectangular mount in a motor bike frame. Photo courtesy of Bruce Roberts*

---

[2] See http://www.agnimotors.com

# Brushed DC series wound motor

The series wound DC motor works in exactly the same way as the brushed Permanent Magnet motor. The only difference is that the magnetic field is created by an electromagnet rather than a permanent magnet. In a series wound motor, the wire creating the field (like the wire coil around the metal bar in Experiment 3 in chapter 2:) is just a few thick turns in series with the rotor (Figure 3-6)

Neglecting losses, this means that the torque produced by a series motor goes up as the square of the current (twice the current, four times the torque, three times the current, nine times the torque).

Here is why. Suppose somehow we just doubled the current in the stator (field) windings and left the rotor (armature) current unchanged. Neglecting losses, this

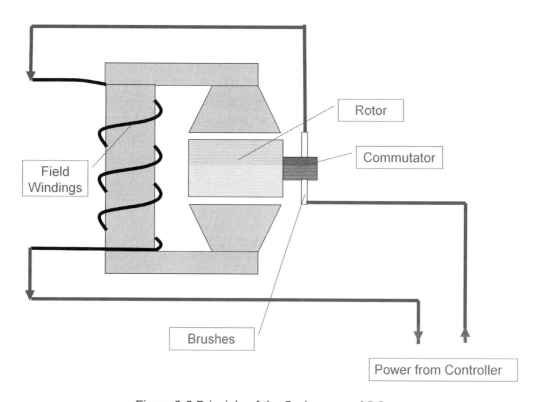

*Figure 3-6 Principle of the Series-wound DC motor*

would double the intensity of the magnetic field which would double the torque on the rotor.

Now suppose instead that we kept the stator (field) current unchanged and doubled the rotor (armature) current. Doubling the rotor current also doubles the torque on the rotor. But in the real world we are doubling the current in both the rotor and the stator at the same time because the field and armature are in series: so doubling the current leads to four times the torque.

This is what makes the series wound motor so formidable in a dragster. An IC engine needs a bit of RPM to make good torque. The series-wound motor produces huge torque at zero revs. How long it goes on producing that maximum torque as the speed rises will depend on the voltage. As the speed rises, the back EMF rises: and because of the high current in the field, it rises quite quickly. A 48 volt milk float will outdrag a sports car for the first ten feet, but after that will get left behind. On the other hand, John Wayland's White Zombie racer with 300+ volts easily humbles most IC engined cars over the standing quarter mile (Figure 3-7). The dual 9" motors in White Zombie deliver 1250 ft. lbs. of torque at zero rpm. This low rpm torque allows White Zombie to out-drag 600 HP IC engined cars. Plus you could drive your granny to the shops in it.

Series wound motors have a couple of drawbacks. Firstly putting over a thousand amps through a set of brushes is a bold enterprise. Brushes will always be a potential weak spot. Secondly a series

*Figure 3-7  White Zombie - a road legal 1972 Datsun. Photo courtesy of Carol Brown*

wound motor will self-destruct if it is ever connected up to a power source directly without a load. This includes, for cars that are not direct drive, applying the throttle with the gearbox in neutral.

Recall that the maximum speed of a motor is reached when the back EMF (the reverse voltage induced in the motor as it turns) is equal to the applied voltage. In an unloaded series-wound motor, the current in the rotor is very low. Because the stator windings are in series with the rotor windings, the current in the stator windings is also low. This in turn means that the magnetic field created by the stator is very weak. A weak magnetic field means lower back EMF. So the chain

of events goes like this:

- Low load, low current
- Low current, weak stator magnetic field
- Weak magnetic field, lower back EMF
- Back EMF below supply voltage, speed increases
- Repeat 1 – 4 until the motor exceeds its maximum rated mechanical speed

One other peculiarity of Series wound motors is that they will run quite happily on AC or DC – they are in fact sometimes called "Universal motors". The reason for this behaviour lies in the series

Figure 3-8 Advanced DC motor.. Photo courtesy of Effran Davis SR

Figure 3-9 Kostov Motor. Photo courtesy of Kostov motors
( http://www.kostov-motors.com)

arrangement. If the polarity of the supply is reversed, both the field and the armature currents are reversed at the same time and the net effect is that the motor continues to rotate in the same direction. This can actually present a bit of a challenge in a direct drive car (i.e. a car with no mechanical gearbox) if you want to allow for reverse. Some series wound motors permit you to reverse the polarity of either the stator or the rotor windings independently, which makes an electrical "reverse gear" possible. At the time of writing, three makes of series DC motor are popular with EV builders:

- Netgain's Warp, Transwarp and Impulse series (the Netgain Impulse 9 is pictured in Figure 3-2)

- Advanced DC (Figure 3-8)

- Kostov (Figure 3-9)

# Other types of DC motor

There are three other types of DC motors that you may come across:

- Shunt Wound

- Compound Wound

- Separately Excited (Sep Ex)

All three are very similar to the series-wound motor: the major difference being the wiring of the stator (field) windings. Shunt and compound wound motors are less applicable to EVs and there are few EVs using separately excited motors, but we will cover these types briefly for completeness.

## Shunt Wound

In shunt wound motors the stator windings are connected in parallel with the rotor rather than being in series with it (Figure 3-10).

In the real world, the low-speed torque of these motors is not as high as that of a similar series motor. They are probably better suited to constant speed applications such as fans and pumps.

## Compound Wound

Compound wound motors are a bit of a throwback to the days before effective solid-state controllers, and are unlikely to be encountered new today, having been superseded by separately excited motors.

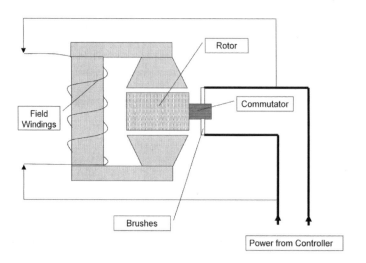

*Figure 3-10 Shunt wound motor schematic*

## Separately Excited

In a "Sepex" motor, the power to the stator (field) is entirely separate from the rotor (armature) supply (Figure 3-11). With modern power electronics it is perfectly possible to control the field current to optimise performance at any combination of speed and torque. In theory, this is the best arrangement, as with a suitable controller almost any combination of "series" and "shunt" characteristics can be dialled in and adjusted by the software to optimise the motor's response. At the time of writing however, suitable controllers are a bit thin on the ground, especially if you are looking for high performance.

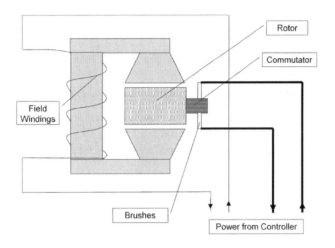

*Figure 3-11 Separately-Excited ("Sepex") Motor*

# AC motors

Both the AC motors that we are going to consider (the induction motor and the so called "Brushless DC" motor) have similar arrangements for the field: in essence a series of electromagnets (or "poles") inside the stator which are energised one after the other to produce a rotating magnetic field, in much the same way that a Mexican wave can appear to move across a stand full of spectators without anything actually changing position (Figure 3-12).

This requires a special controller which rhythmically increases and decreases the currents in the pole field windings to produce the rotating magnetic field.

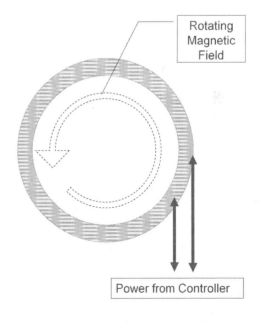

*Figure 3-12 Rotating Magnetic field inside an AC motor*

The bulk of the electrical power is fed into the (fixed, stationary) field windings. There is usually no power supplied directly to the rotor. Note that this arrangement has a huge advantage over the DC motors that we have been looking at – *no brushes*

# AC Induction motors

You will recall that in Experiment 1 in Chapter 2 we moved a wire in an electric field and it generated a current in the wire. Imagine a loop of wire inserted into the rotating magnetic field generated by the field windings described above. The movement generates a voltage in the wire which will (if the circuit is complete) cause a current to flow. Also, from Experiment 4, a current-carrying wire in a magnetic field experiences a force on it. The end result is that a simple loop of wire on a pivot placed in the rotating magnetic field will be "dragged round" electrically. You can think of it as being like dipping a spoon into an open jar of honey sitting on a turntable (Figure 3-13). It is important to note that the rotor windings do not have to be connected to anything (although they are in some cases because this can improve control of the motor). The simplest induction motors are sometimes called "squirrel cage" motors because the rotor consists of nothing more than some metal strips or bars arranged like a pet's exercise wheel (Figure 3-14). Others use more conventional wire windings.

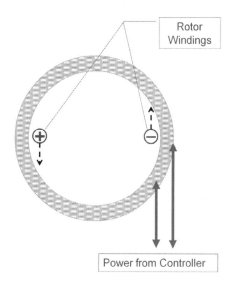

*Figure 3-13 AC Induction Motor*

*Figure 3-14 Induction "Squirrel Cage"*

*Figure 3-15 Tesla Roadster. Photo courtesy of David Breach*

*(http://www.flickr.com/people/davidbreach)*

In any form of induction motor the rotor never quite keeps up with the rotating magnetic field. There is always a bit of "slip" i.e. a speed difference between the rotating field and the rotor. If the rotor was turning at the same speed as the magnetic field, there would of course be no relative movement, so no induced current in the rotor, so no torque. This means that the induction motor is a kind of *asynchronous* motor.

Most of the motors connected to mains electric power around the world are AC induction motors. They are about as simple and robust as you could hope to find, and for single speed use from the mains they do not need any form of controller. The Tesla Roadster (Figure 3-15) and Tzero sports cars both use induction motors manufactured by AC Propulsion.

There are a number of other AC motors sold for use in EVs (see for example Figure 3-16, Figure 3-17 and Figure 3-18). These tend to be somewhat more expensive than their DC counterparts.

Figure 3-16 MES DEA. Photo
courtesy of MES s.a.
(http://www.mes.ch)

Figure 3-17 AC55 and DMOC 445.
Photo courtesy of Azure Dynamics
Force Drive
(http://www.azuredynamics.com)

Figure 3-18 Brusa HSM 1-6.17.12 motor
and transaxle gearset. Photo courtesy of
BRUSA Elektronik AG
(http://www.brusa.biz)

# Brushless DC (BLDC) Motors

Despite their name, these have more in common with AC motors than DC ones. They have broadly the same kind of rotating field as the induction motor, but the rotor is equipped with permanent magnets – the magnetic interaction between the permanent magnets and the rotating magnetic field is what generates the torque (Figure 3-19).

The rotor in a brushless motor normally remains in lock-step with the rotating field. Brushless DC motors are therefore said to be *synchronous* i.e. there is no slip in normal operation.

There is commonly (but not always) another difference between Brushless DC and AC induction motors. An AC Induction motor power

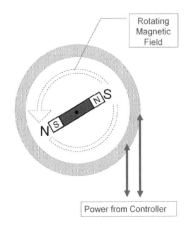

*Figure 3-19 The Principle of a Brushless DC motor*

supply is "sinusoidal"; that is the voltage and current rises and falls smoothly from zero to maximum positive voltage then back to zero and on to maximum negative voltage, then smoothly back to zero again (Figure 3-20).

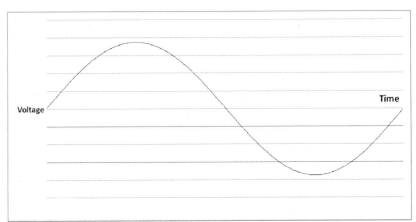

*Figure 3-20 AC power supply (one pole)*

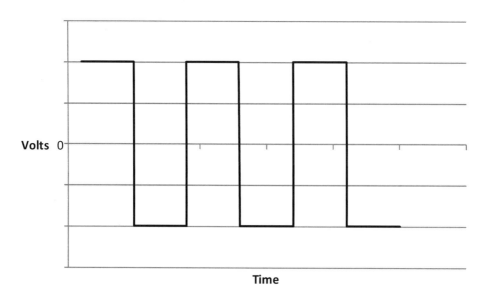

*Figure 3-21 Brushless DC Power Supply (one pole)*

Brushless DC motors are by contrast frequently designed to operate with a simpler cyclic on-off switching. One part of the stator windings are energised, then that part is switched off and the next one energised. This is of course precisely what a commutator does (apply voltage to one winding segment, then switch off that one and move on to the next) - Figure 3-21.

So why are these motors called "Brushless DC"? At one level you can think of them as being just like the brushed permanent magnet motor we described earlier but turned inside out: instead of a commutator energising *rotor* windings in sequence, the controller for a Brushless DC motor energises *stator* windings in sequence. The brushed DC motors has permanent magnets in its stator. The Brushless DC motor has permanent magnets in its rotor.

|  | Brushed Permanent Magnet | Brushless DC |
|---|---|---|
| Stator | Permanent Magnet | Sequential switching of current in stator by the controller |
| Rotor | Sequential switching of current in rotor by the commutator | Permanent Magnet |

*Figure 3-22 M & C Brushless DC Bus motor Weight 350kg, Power 420 kW, Torque 1kN-m. Photos courtesy Guangdong M&C Electric Power*

*Figure 3-23 M&C Brushless DC Bicycle Hub Motor, Weight 2.9 kg, Power 180 Watts, Torque 6.5 N-m*

There are a number of manufacturers building Brushless DC motors. For example M & C motors in China manufacture a range of these motors for use in everything from buses (Figure 3-22) to electric bikes (Figure 3-23).

# Other types of AC motor

## Reluctance motors

You may recall from chapter 2 that a magnetic field passes through iron far more easily than air. One of the side-effects of this phenomenon is that a magnetic field will exert a force on iron to minimise the amount of air that it has to pass through. This is the basis of the solenoid.

There is also one form of electric motor which exploits this behaviour. The *Reluctance* motor offers a magnetic field a nice easy path through lots of iron rather than all that nasty air: all the magnetic field has to do is just rotate the armature a little bit until things line up nicely...

Reluctance motors aren't used much, if at all, for EV propulsion but you might find them driving ancillary equipment. Stepper motors are typically a form of reluctance motor.

## Wheel motors

Most EV conversions use an electric motor as a replacement for the IC engine and still require a mechanical transmission (gearbox, differential, half-shafts: and sometimes a propshaft).

It is in however possible in principle to use motors which fit inside a wheel hub, eliminating the need for these mechanical components. The idea of individual in-wheel motors driving each wheel of a vehicle is not new. In 1900 at the World's Fair in Paris the 'System Lohner-Porsche' was debuted. This vehicle was designed by Ferdinand Porsche; the man who gives his name to the Porsche car company. His first job in the automotive world was with Jacob Lohner. The 'System Lohner-Porsche' was an electric vehicle driven by two in-wheel motors; this vehicle was capable of over 35mph and set several Austrian speed records. (Figure 3-24)

*Figure 3-24 Lohner Porsche of 1900 driven by in-wheel motors. Photo © Porsche AG*

There are organisations working on in-wheel motors today. For example Protean Electric (formerly PML Flightlink) has demonstrated their Protean Drive in-wheel motors in a variety of different vehicles and is currently working with a number of different car manufacturers and suppliers to bring this technology into volume production (Figure 3-25).

*Figure 3-25 Protean Wheel Motor. Photo courtesy of Protean Electric[3]*

This configuration has the massive advantage that no transmission is needed. On the down side, electric power has to cross the gap between chassis and suspension. Also, if the motor is direct drive, the motor RPM is low which means that the torque has to be greater for a given power output.

There is also a potential issue with unsprung weight. One of the targets of car designers is to minimise the weight of wheels, brakes, tyres and suspension

---

[3] http://www.proteanelectric.com

components to improve ride and handling. Wheel motors do not help here, although the importance of this is often overplayed: a 2010 report (*Unsprung Mass with In-Wheel Motors - Myths and Realities*, Anderson M, Harty D presented at AVEC 2010) concluded that "perceptible differences ... are small and unlikely to be apparent to an average driver."

# Motor ratings and motor cooling

Compared with an IC engine, electric motors are extremely efficient. But they are not 100% efficient and they are also physically small. A motor putting out 30 kW with 90% efficiency is still generating 3 kW of heat in a space little larger than a big water melon. A 3kW electric heater soon warms up a living room, so without adequate cooling, 3kW will generate high temperatures inside a motor.

Most EV motors are air cooled. An obvious solution is to put a fan on the motor shaft (most electric tools take this approach for example). This is OK, but it does mean that cooling air flow drops off at low speed which can be an issue. Brushed DC motors probably have greater cooling problems, because most of the heat is generated in the rotor deep inside the motor.

One thing that electric motors do have going for them is their mass. They contain a lot of metal and in consequence heat up slowly and cool down slowly. This means that they are capable of providing short bursts of higher power than they are capable of sustaining continuously.

Motors typically have a continuous rating (maximum current) and maybe one or more short-period ratings. This fits quite well with the profile of use in a car: a maximum-effort acceleration with the motor drawing (say) 1000 amps tends to last no more than 10 or 15 seconds for the simple reason that accelerating that hard for any longer will put you over the legal speed limit and/or into a regime where back EMF will start to reduce the current anyway.

All the same heat is an important consideration, and one you ignore at your peril

# In Conclusion

This completes our gallop through electric motors as applied to EVs. If you would like to go further with your understanding of motors, *Electric Motors and Drives: Fundamentals, Types and Applications* by Austin Hughes (ISBN 978-0750647182) is an excellent introduction at the next level of detail. In the meantime we'll move on to controllers: the boxes packed with power electronics which enable us to make use of electric motors.

# 4: Controllers

Controllers are linked directly to the accelerator (throttle) pedal of an EV. They occupy exactly the same niche as a carburettor or fuel injection system in an IC-engined car; that is, they allow the driver to control power and hence speed (Figure 4-1). Just like a carburettor or fuel injection system, a controller needs to be matched to the motor. A puny carburettor on a big IC engine will strangle it. A puny controller will likewise strangle a powerful motor and good batteries. Controllers also need to suit the type of motor that they are feeding: a controller designed for a DC motor for example could not even make an AC induction motor turn over under no load.

In the context of EVs, Controllers can be

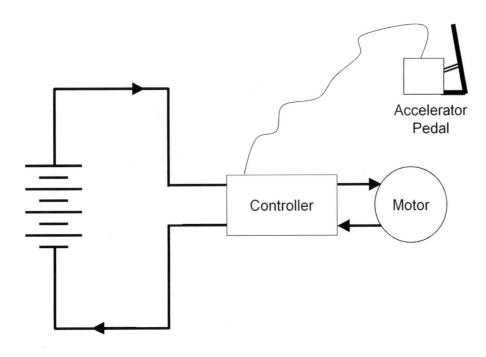

*Figure 4-1 Battery, Controller and Motor*

placed into one of four groups as follows:

- Controllers for Separately Excited DC motors
- Controllers for other brushed DC motors
- AC Induction Controllers
- Brushless DC (BLDC) Motor controllers

A controller for a separately excited motor might work with other DC motors, and an AC induction controller might work (very imperfectly) with a brushless DC motor, but none of these types of controller are otherwise interchangeable: indeed an AC Induction or Brushless DC system is best acquired as a matched motor/controller package. A controller designed for one make and model of AC Induction or BLDC motor might not work well even with another motor of the same type.

The simplest controller is the type typically used with a Series wound DC or a brushed Permanent magnet motor. We'll start with that.

## The simple DC motor controller

The very simplest form of DC motor controller (never used in an EV) is a variable resistor (or rheostat) in series with the motor. Increasing the resistance reduces the current flow, which in turn reduces the power developed by the motor (Figure 4-2). It's as simple as that.

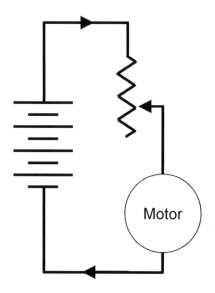

*Figure 4-2 Crude Motor Control with Variable Resistor*

This approach works well for tiny motors in models and toys, but it would be totally impracticable in an EV. The problem is that much of the available power is burned up in the resistor. This slashes range, and also creates a heat problem. All that waste energy has to go somewhere, so the variable resistance would need to be physically large and well-cooled or it would burn out.

As it turns out there is a solution to the DC controller problem which is even simpler conceptually. Suppose that you were presented with an EV that had no controller

– merely a big on-off switch. Could you drive it across town? Well yes you probably could, although you would be unwise to try it, and it would be an uncomfortable trip. When you wanted to move off you would switch on, wait until you had built up enough speed and then switch off and coast. As your speed dropped you would switch on again until you were back to your target speed. You would progress down the highway in a series of pulses: switch on – accelerate – switch off – coast – switch on – accelerate....

DC controllers use this principle but immensely speeded up. They generate a series of full-voltage pulses interspersed with periods of no voltage. The ratio of "on" time to "off" time governs the average voltage and therefore the power; and the beauty of it is that there is little wasted energy.

If you were to measure the voltage output of a typical DC controller with a 100 Volt supply at a high power setting a little short of full power it would look like Figure 4-3. On the other hand controller output at a low power setting might look more like Figure 4-4

This is known as Pulse Width Modulation or PWM. The frequency is typically high:

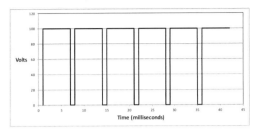

*Figure 4-3 High Average Voltage*

thousands of pulses per second. Ideally the frequency is high enough to be beyond the range of human hearing (it isn't always that high though – some EVs emit an audible whine).

So much for the voltage. The resulting current is a bit different. One of the peculiarities of electric motors (and many other electrical devices) is that there is something like a flywheel effect on any electric current flowing through it. The current takes a little while to build up when a voltage is applied, and a little while to die away when the voltage is removed. It is as if the current has a kind of inertia and is reluctant to increase or decrease. It isn't really physical inertia but from the outside it

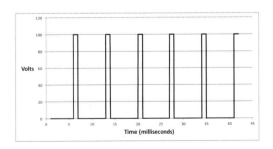

*Figure 4-4 Low Average Voltage*

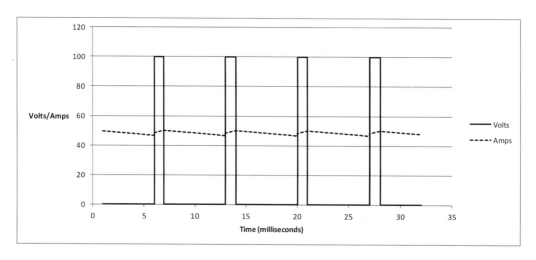

*Figure 4-5 Motor current and voltage*

appears to behave in a similar way. This characteristic is described as **inductance** and it can be measured, just as resistance can.

One of the effects of inductance in the motor is that at these high frequencies the current through the motor varies much less than the voltage: and the higher the frequency, the less it varies. Figure 4-5 illustrates typical current variation with voltage changes.

This means that the controller has to be designed to allow for current flow continuing even during that part of the cycle where there is no supply voltage. If it behaved like a mechanical switch and just interrupted the current on every cycle, there would be mayhem: sparks, spikes and burned out

components. A path for continuing current flow is therefore created using so-called "free-wheeling diodes". More on this later.

So how is the pulse width modulation achieved? You can think of a DC controller as having two halves – the brain and the brawn. The "brain" is basically a pulse generator. It translates the throttle position into a series of low-power pulses where (as in Figure 4-3 and Figure 4-4) the proportion of "on" time to "off" time depends upon the throttle position. The "brawn" switches the main drive voltage in time to the signal coming from the pulse generator (Figure 4-7).

The "brawn" consists of one or more power transistors - solid state electronic switches. The application of a low voltage at one

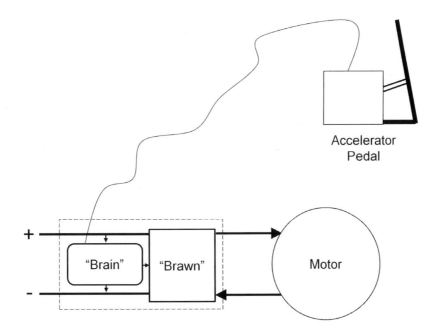

Accelerator
Pedal

"Brain"   "Brawn"   Motor

+

−

*Figure 4-7 DC Controller High Level Schematic*

terminal of the "switch" turns it on: removal of the voltage turns it off (Figure 4-6).

This type of controller has only been practicable for a few decades, since the advent of "power electronics" - solid state components (primarily different forms of transistor) that can handle large currents.

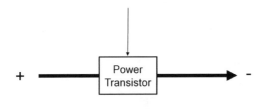

+   Power Transistor   −

*Figure 4-6 Schematic of a Power Transistor*

Even today, big power transistors are expensive in high power ratings. As a result, controllers are not cheap, and the price is linked to the maximum current and voltage ratings of the controller.

So what can go wrong with this kind of controller? You are entitled to assume that reputable manufacturers of controllers know what they are doing, and indeed anecdotal evidence suggests that controllers are pretty reliable. The *White Zombie* team (*White Zombie* is the street-legal electric dragster referred to earlier) has damaged motors and batteries in pursuit of shorter and shorter standing ¼ mile times – but has rarely reported controller problems.

There are a couple of things to be aware of though. The first is a safety issue which you are unlikely to encounter with a good modern system, but could still be a problem. Some controller designs will go to full power if the wires connecting the throttle to the controller break or become disconnected. This is an obvious safety hazard, akin to the throttle of an IC engined car jamming wide open.

The second issue to be aware of is heat. The best of controllers are not quite 100% efficient and so do produce waste heat. If the controller is pulsing 1000 amps at 300 volts (an extreme, but possible scenario) and is 99.5% efficient, it will still need to dissipate 1.5kW – as much output as some mains operated space heaters. The critical issue here is the speed with which the power electronics can switch from fully off to fully on. When the switching device is fully on, there is little or no voltage drop across it, so little or no power is lost even if it is passing hundreds of amps (power = volts x amps so if either is zero, power is zero). When it is fully off the switching device itself carries no current, and once again there is little or no power lost even if there is a high voltage across the switching device. The problem is the tiny instant of time in each cycle that it is partially on. The power dissipated can be very large albeit for a very short time; and it is generated inside the large and expensive semi-conductors that do the switching. So be lavish in your cooling arrangements for a high power controller. If it is water cooled, ensure it gets what it needs. If it is air cooled, don't shove it in some corner where the air cannot circulate.

That is about it for the simplest DC controllers. We'll now move on to controllers for separately excited DC motors.

# Separately excited DC

You will perhaps recall from Chapter 3 that separately excited motors require a controller which can adjust the voltage applied to the field (stator) windings quite separately from the supply to the armature (rotor) – see Figure 4-8.

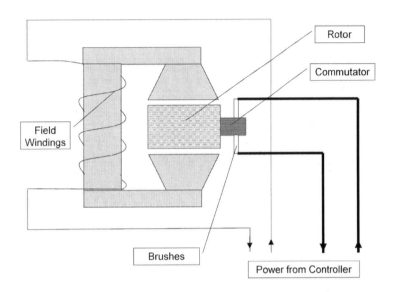

*Figure 4-8 Separately Excited Motor schematic. Note separate connections for field and armature*

The major difference between the simple DC motor controller described earlier and the controller for the separately excited motor is therefore that the latter must control two outputs – one each for the stator (field) and the rotor (armature) windings.

At low speeds, the field current can be set to optimise the strength of the magnetic field in the motor to give maximum torque potential. At higher speeds however, something quite unexpected begins to happen. The controller actually reduces the field current, reducing the strength of the magnetic field in the motor.

This is counter-intuitive. Why does weakening the magnetic field allow the motor to run faster? The answer to this conundrum is related to the back EMF generated in the rotor. You will recall that there are two opposing voltages at work in a motor – the external applied voltage from the controller, and a voltage in the opposite direction resulting from the generator action that happens in any motor (see Back EMF in Chapter 2:Refresher course).

The current flowing in the rotor is proportional to the difference between the applied voltage and the back EMF. An electric motor reaches its maximum speed when this net voltage (Applied voltage minus back EMF) gives just enough current to keep the motor spinning. The stronger the magnetic field in the motor, the higher the back EMF: or to put it another way, the stronger the field the lower the speed at which this critical back EMF is reached. Weakening the field reduces the back EMF, allowing an increase in speed.

The weaker field does however reduce the maximum torque available from the motor, but therein lies the benefit of the Sepex motor and controller: high torque at low speed like the Series wound DC motor, combined with high speed albeit with lower torque.

# AC Induction Controllers

Controllers for AC induction motors are, as we shall see, clever gadgets. They can not only rub their tummy and pat their head at the same time, they can do this whilst simultaneously playing the Hallelujah Chorus on a mouth organ and riding a unicycle down a mountain. It is fortunately a lot easier to understand the principle of operation of an AC induction controller than it is to design one.

## Mains-operated AC Induction

Probably the best place to start is to think about the AC Induction motor hooked up to a three phase mains supply. There are millions of them in factories and workshops all over the world.

Household electricity supply consists of a single "live" delivering a straightforward AC voltage. The plot of voltage against time looks something like Figure 4-9

On the other hand, commercial premises are usually offered three phase power – that is three "live" conductors each carrying an AC voltage. Critically, the three voltages are not in step with each other: each is shifted a third of a cycle from the previous one so that a graph of voltage against time look like Figure 4-10.

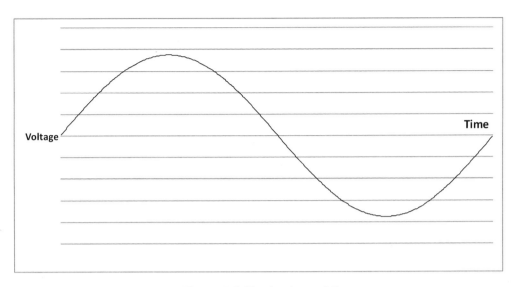

*Figure 4-9 Single phase AC*

This has some very happy consequences. Firstly, the three phases sum to zero which means there is very little current in the common neutral or return conductor. This greatly simplifies distribution.

Secondly if you hook each of the three phases up to successive "Poles" (segments of the field windings) in an AC Induction motor, you get the rotating magnetic field

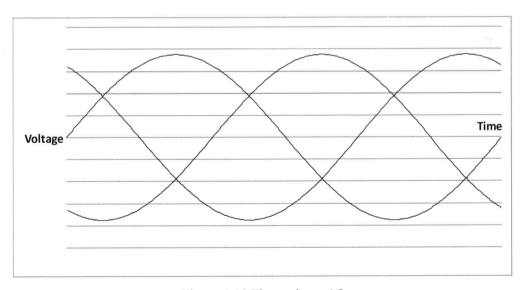

*Figure 4-10 Three phase AC*

you need and the motor will run with good efficiency with no controller whatever: albeit with no speed control to speak of. There is nothing in the entire discipline of engineering that is more elegant, unless perhaps it be the stone arch (stone is good in compression, lousy in tension or bending. A stone arch bridges a wide gap using just stone in compression. Who says there is no poetry in science and technology?).

You will recall from the section on AC induction motors that they are *asynchronous* – that is, the rotor rpm is always just a little lower than that of the rotating magnetic field. You might also recall that the difference is known as the *slip* (Figure 4-11)

As the load on the motor increases, the speed of a well-designed AC Induction

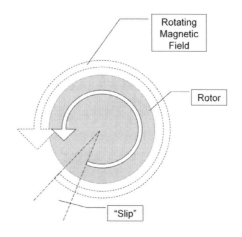

*Figure 4-11 "Slip"; the difference between the speed of rotation of the field and of the rotor*

motor will initially decrease but not very much. A mains-connected induction motor in its design operating range is an almost constant speed device. As the load increases the speed drops very slightly, increasing the slip, and the motor just sucks more amps.

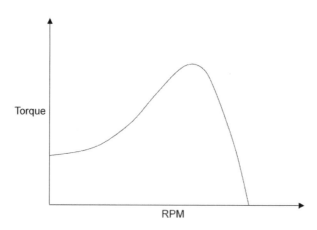

*Figure 4-12 Torque/Speed Curve for AC Induction Motor at Constant Frequency*

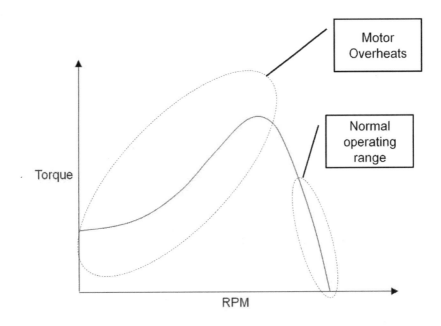

*Figure 4-13 Normal operating range of AC induction motor*

Up to a point. As the load continues to rise the slip becomes so great that the torque produced by the motor peaks and starts to drop again. Unless the load drops off very rapidly as the speed drops, the motor stalls. The speed torque curve looks like Figure 4-12

This is very bad news for anyone hoping to use a simple variable voltage to control an AC induction motor. Remember that the Series-wound DC motor produces maximum torque at zero revs? This is not the case with most AC induction motors operating at constant frequency. Dropping the voltage will increase the slip slightly at a given torque output, but a car using an AC Induction motor with a constant frequency controller would be a gutless wonder away from the lights and would only be efficient and effective in a narrow speed band. Furthermore the motor would overheat badly at low speeds (Figure 4-13)

There are two ways to address this problem. The first is a workaround which is common in industry but which to my knowledge is never used in an EV. The second is the norm for controllers used in EVs.

## AC Induction – compromise

It is possible to design an AC induction motor so that it produces maximum torque at zero rpm, which would permit voltage control of the motor speed. Unfortunately, this is achieved by adding resistance to the stator windings! This is highly undesirable, having, as it does, at least three negative effects. It;

- reduces the efficiency of the motor/controller combination, thus reducing range in an EV

- creates additional heat which must be dissipated

- reduces torque at higher rpm, thus reducing acceleration

This is however the approach used in many industrial environments. The torque curve shown in Figure 4-14 is typical of a particular class of three phase industrial motors (driven from the mains without a controller) in applications that need a high starting torque.

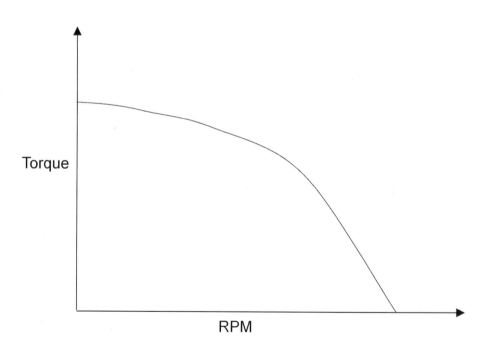

*Figure 4-14 AC Induction Motor with resistive windings*

## Variable frequency AC

The most efficient way to control an AC induction motor is to use a variable frequency drive. Look at the constant frequency torque/speed curve depicted in Figure 4-12. Now imagine that curve replicated at a number of lower frequencies as in Figure 4-15.

This is easy to describe, but not so easy to do, particularly when you are handling hundreds of amps. And there is a further complication. The field windings of an AC motor are highly inductive (you may recall that inductance is a measure of the ease with which you can change or reverse an electrical current in a device). This means that the voltage needed to achieve a given current varies with frequency. So to adjust the motor speed, the controller not only has to alter its output *frequency*, it also has to adjust its output *voltage*.

Most controllers achieve this by using a sophisticated form of pulse width modulation (PWM) as described in the section on DC motor controllers (see for example Figure 4-3 and Figure 4-4). To understand how this works, we'll conduct a mental experiment.

Imagine for a moment that you were driving an EV equipped with an ordinary DC motor and controller of the kind we described earlier down a straight but deserted road. Imagine that you pushed the throttle pedal

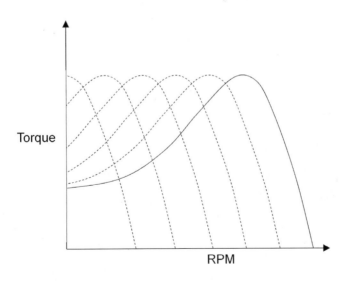

*Figure 4-15 Torque/Speed at Various Frequencies*

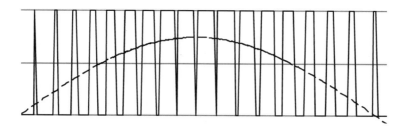

*Figure 4-16 DC Controller response to opening then closing throttle*

slowly down to near maximum and then let it come up again. You might get an output from the controller that looked something like Figure 4-16.

At the start and end of your experiment, the throttle would be nearly closed and the pulses from the controller would be short. As the throttle opened, the pulses would get wider and wider until near full throttle they would almost merge. The dotted line in Figure 4-16 is the average voltage that the motor sees. Clearly, the faster you cycled the throttle the shorter and steeper the curve of resultant voltage would be.

If this "resultant voltage" curve looks uncannily like half an AC cycle, that is because it is half an AC cycle. An AC Controller uses exactly this strategy to mimic an AC waveform (Figure 4-17)

By varying the timing and duration of the pulses, the AC controller can produce an output waveform that is equivalent to any desired frequency and voltage.

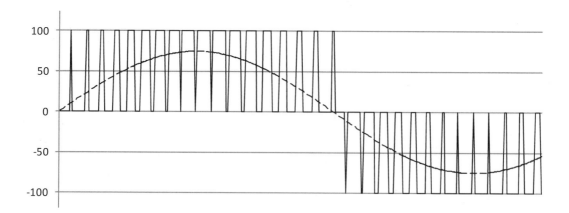

*Figure 4-17 AC waveform synthesised from DC pulses*

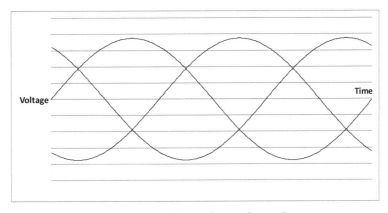

*Figure 4-18 Resultant Three Phase Output*

Or rather, any three waveforms. We have been considering just a single pole in a three pole machine. A real controller has to do everything we have just described, but do it three times, whilst keeping the phase difference between the resulting waveforms just right (Figure 4-18).

This brings us back to the three phase mains power where we started: but unlike the mains, the AC motor controller can adjust the frequency and voltage of all three phases at will.

So how does it do it? Broadly speaking the AC controller, like the DC controller, has two core modules: low power electronics that create a complex pattern of low energy pulses used to control hefty (but comparatively simple) electronic switches (Figure 4-19). A basic DC controller needs just one wire between these two modules, but the AC controller needs six; and there are three wires (one for each phase) running between the power electronics and

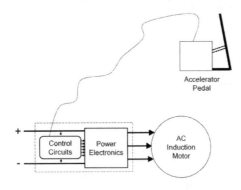

*Figure 4-19 Block Diagram of an AC Controller*

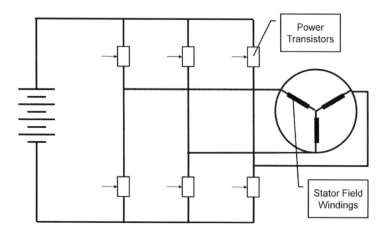

*Figure 4-20 AC Induction Controller Power Module*

the motor.

Similarly, the power electronics module of a DC controller need contain only one power transistor but the AC Controller needs six: two per phase. One half of each pair handles the positive half of the AC cycle, and the other one handles the negative half (Figure 4-20). As before there are free-wheeling diodes (not shown).

The control circuit is the clever bit. Whilst it would be possible to build a control circuit of this kind from discrete electronic components there are off-the-shelf integrated circuits available commercially which greatly simplify this task.

## Vector Control

The arrangements described so far (known as Scalar Control) are adequate for many purposes. However, further improvements in the accuracy of control can be made by feeding information about various motor parameters (such as rotor position or field currents) to the controller. This further complicates the control algorithm, but improves the low r.p.m. performance, and also response to sudden changes in power demand (Figure 4-21)

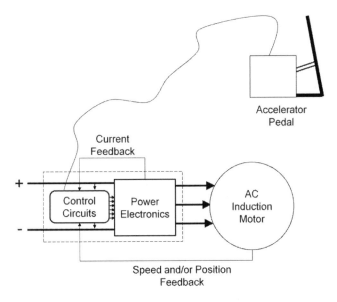

*Figure 4-21 Vector Control Block diagram*

# The Brushless DC (BLDC) controller

Brushless DC (BLDC) motors are unlike AC Induction motors in two respects. Firstly, BLDC motors are synchronous; that is they run with zero slip. A BLDC motor is either running at synchronous speed or it is stalled. There is no half way house. Secondly BLDC motors do not need a sine wave power supply. Power to the motor poles needs merely to be switched on and off in the correct sequence.

The best way to understand controllers for BLDC motors is to start by thinking about the ordinary (brushed) DC Permanent Magnet motor. You can think of the commutator in a brushed motor as being a position-sensitive switch. The commutator switches the supply current to the *rotor* windings in sequence so that they produce maximum torque as they interact with the field set up by the permanent magnets on the *stator*.

The BLDC motor works in just the same way but rotor and stator are the other way around, so that the controller switches the supply current to the *stator* windings in sequence so that they produce maximum torque as they interact with the field set up by the permanent magnets on the *rotor*. It

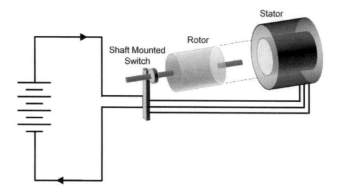

*Figure 4-22 Theoretical BLDC mechanical controller*

would be perfectly possible in theory to use a cam-operated switch on the shaft of a BLDC motor to switch power between the stator windings (Figure 4-22).

The theoretical BLDC motor pictured in Figure 4-22 could be made to work: but many of the advantages of reliability would have been lost: we would merely have substituted one mechanical device (the brushes) with another

(the shaft-mounted switch).

Real BLDC controllers use all electronic solid state components with no moving parts. The classic layout is to use a solid-state position sensor on the motor shaft to control a power electronics module of virtually the same layout as the one used by AC Induction Motor Controllers discussed earlier (Figure 4-23).

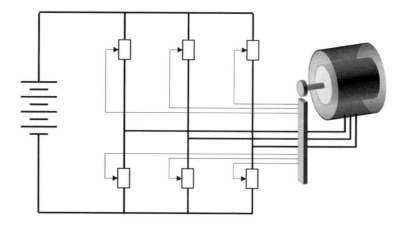

*Figure 4-23 BLDC controller block diagram*

Once this switching arrangement is in place, speed/torque control of a BLDC motor becomes simply a matter of using pulse width modulation of the DC in just the same way as for the brushed DC motor.

## Sensorless Control

There is a variant of the BLDC controller that even dispenses with the solid state position sensor. So-called "sensorless" controllers use the back EMF generated in the field windings that are not energised. This reduces both motor cost (by eliminating the position sensor), and wiring complexity. Special arrangements must be made for starting from rest and for very low speed operation when there is insufficient back EMF.

# Regenerative Braking

In most IC-engined cars when you take your foot off the throttle on the move, the engine operates as a brake (this is not true of all cars – some Rovers and early Saabs among others were fitted with a freewheel to conserve fuel: when you took your foot off the throttle the engine rpm dropped to idle and the car just coasted).

The simplest electrical drivetrains do not exhibit "engine braking". When you take your foot off the throttle, the vehicle just coasts and must be slowed using conventional brakes. However with most types of motor it is possible to replicate IC engine braking behaviour. When it is done correctly, this form of braking recovers some electric power – this reduces the demands on the conventional braking system and also extends range slightly.

The technique relies on the fact that an electric motor is, in most cases, electrically identical to a generator: indeed an electric motor that is being turned faster than its no-load speed IS usually a generator. If current is drawn from a motor in this state, the effect will be to create a braking torque as the energy stored in the rotating motor and load is converted into electrical power.

The simplest example of this effect is short-circuiting the rotor windings in a rotating DC motor. This generates a high current in the rotor in the opposite direction to the normal motor current. This in turn creates a braking torque. The braking energy is dissipated as heat. This crude approach is effective, but the preferred technique in EV drivetrains is to use the power generated by braking to recharge the battery rather than to throw it away as heat. This recharging approach is referred to as "regenerative braking". Controllers that allow for regenerative braking may also allow adjustment of the amount of regeneration depending upon throttle position and/or the amount of conventional braking going on.

Regenerative braking is not a magic bullet. Experience suggests that range improvement with "regen" may be of the order of 10% and there is even anecdotal evidence that regeneration can trigger a change in driving style which actually reduces efficiency. In some cases, a 10% increase in battery capacity may be a cheaper and simpler alternative.

There are several issues associated with regenerative braking:

- 1, 2 and 4 quadrant operation
- Overcharging
- Battery C rating

## 1, 2 and 4 quadrant operation

An ideal motor/controller combination can operate as a motor or as a generator, running in either direction. This is referred to as four-quadrant operation (Figure 4-24):

Some motor/controller combinations cannot manage full four-quadrant operation: they might allow just motoring in a forward direction (single quadrant) or two-quadrant (forward and reverse but no generator action).

The simplest series wound DC motors are inherently single-quadrant devices: typically they are either incapable of regeneration, or implement it poorly. In some cases they cannot be reversed either. To reverse a DC motor, either the magnetic field in the motor must be reversed or the current in the rotor (armature) must be reversed – but not both.

|  | Motor | Generator |
|---|---|---|
| Forward | Q1 | Q3 |
| Reverse | Q2 | Q4 |

*Figure 4-24 The four quadrants of operation of a motor/generator*

In a series wound DC motor, the field windings are (by definition) in series with the armature so reversing the current in one reverses the current in the other. So simply reversing the polarity of the supply to such a motor makes no difference to the direction of rotation. In practice, such motors often have separately-accessible contacts for the field windings. This allows the use of a reversing contactor.

Series wound DC motors have another issue with regen. When there is no current in the motor there is no magnetic field (or very little – maybe just a weak residual field left over from the last time there was a current in the motor); no magnetic field means no voltage is generated in the rotor windings, so no current flows, so there is no braking effect. There are also potential issues with the positioning of brushes to avoid excessive sparking.

## Overcharging

Normally an EV battery should have no trouble storing the energy created by regenerative braking. After all, the energy to accelerate the car came from the battery in the first place. However, there could in theory be a problem if a battery that was already fully charged was asked to absorb further energy from regeneration. This could (again hypothetically) be an issue if – say – a driver used regeneration in a long descent just after setting off. It could also be an issue with hybrids (HEVs) if such a long descent occurred just after recharge from the IC engine had completed. Note however that 30 seconds of braking at 100 Amps would represent less than 1% of the capacity of a 100 Amp-hour battery.

## Battery C rating

In the section on batteries we discuss battery C rating: the ratio of maximum allowable battery current to battery amp-hour capacity. In that context we are concerned with C rating on discharge – how fast you can extract power from the battery to accelerate the vehicle. Batteries (particularly Lithium ion batteries) may also have a maximum C rate for charge. The maximum C rating for charging may be lower than the maximum discharge C rating: in other words you cannot safely push power back into the battery as quickly as you can pull it out. This means that the maximum regen current (and thus the maximum regen braking torque) may be limited by the battery.

# Some Typical Controllers

This section shows some typical controllers. These are not the only ones available, nor even necessarily the best. They do however give a flavour of the kind of thing that is available commercially at the time of writing. The specifications were current at the time of writing, but may be subject to change.

## Curtis 1231C-86XX

Series Wound or Permanent Magnet DC

96-144 Volts, 500 Amps (2 minute rating), 72 kW

Curtis controllers are popular. At the time of writing they are readily available from a number of sources at reasonable prices. Power output is modest so they are better suited to a light road vehicle with a gearbox rather than, say, a drag racer.

Photo courtesy of
Curtis Instruments, Inc.
(http://curtisinstruments.com)

## Zapi SEM3

Separately excited DC

48 - 96 Volts, 400 Amps (600 for 48 Volt version), 38 kW (theoretical) – 12 kW (recommended)

This Zapi controller supports regeneration. (In this photo, the Zapi controller is the oblong box mounted over the top of the motor and behind the fan).

Photo courtesy of
Nikos Giakoumelos

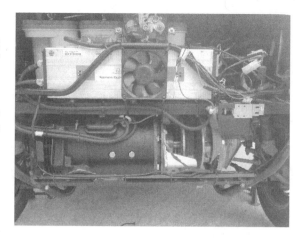

## Zilla Z2K

High Performance DC

72 – 300 Volts, 2000 Amps,

600 kW (Theoretical)

The controller of choice for electric dragsters and the like. Note that at the time of writing future production of Zilla controllers was uncertain.

Photo courtesy of
Kurt Clayton
(http://kurtschevys10ev.blogspot.com)

## Curtis 1238-75XX

AC Induction

72-96 Volts nominal, 550 Amps (2 Minutes), 52 kW (Theoretical)

Controllers in this family are sometimes sold as a package with the popular AC50 AC Induction motor

Photo courtesy of
Curtis Instruments, Inc.
(http://curtisinstruments.com)

# In Conclusion

The controller is a major element in an electric drivetrain. A powerful controller may be more expensive than the motor it controls. Controller choice has a significant impact on the performance, efficiency and drivability of the finished vehicle. It is as important as the choice of battery or motor. Indeed, these three items (battery, motor, controller) should be chosen to complement each other. Using a Zilla Z2K with a low voltage, low C rating battery wastes the Zilla's prodigious capacity. Using a small Curtis controller with a 13" DC motor is likewise pointless, as the performance potential of the big motor cannot be delivered by a controller limited to around 500 Amps.

So far we have looked at motors and controllers. In the next chapter we turn to the third main element in the EV drivetrain: the propulsion batteries.

# 5: Batteries

If the motor is the heart of an EV or HEV then the batteries are its Achilles heel. The limitations of batteries are the main reason for the tiny proportion of EVs on the road. You have to address at least the following issues when specifying the batteries for your car:

## Energy Density

Batteries store a pitiful amount of energy per unit of volume (energy density) and per unit of mass (specific energy) compared with petrol or diesel fuel. One consequence of this limitation is that you need a lot of batteries to get further than the end of your road. The battery pack is usually the heaviest, bulkiest and most expensive element in a conversion. Inevitably the best batteries in terms of specific energy also tend to be the most expensive. More on this later.

## Charge rate

Some batteries can only be recharged at a limited rate, so you cannot always get back on the road as rapidly as you might like after a trip which largely discharges the batteries.

## Power Density

Power density is the rate that you can draw power (measured in kilowatts or horsepower) from the batteries per unit of volume. Specific power is the equivalent per unit of weight. Batteries have limits on the maximum discharge rate, i.e. a limit on the current (and thus power) that you should attempt to draw. This can limit acceleration: even if your motor and controller could handle more amps, your batteries may not be able to provide them. We touch on specific power later.

## Depth of discharge

If you routinely run your batteries completely flat, you may (with some chemistries) shorten their life. Discharging down to 70% or 80% of maximum capacity is probably a good compromise with many types of cell.

## Cell imbalance

A vehicle battery pack is made up of (typically) 24 – 150 individual cells. Particularly with some battery chemistries, these cells may get out of step with each other so that if you charge the pack as a whole, some cells may be overcharged and others undercharged.

## Maintenance

Some types of battery need routine maintenance such as watering. This can be time consuming and messy.

## Safety

Conventional fuels can cause fire and explosion, but a high voltage battery pack can kill you just as dead (albeit with a lot less drama). Some battery types catch fire when abused and even the humble lead-acid battery can explode if overcharging causes venting of hydrogen and there is a source of ignition in the vicinity.

## Cycle life

Batteries have a limit on the number of times they can be charged and discharged. Beyond this limit either their performance will have degraded to unacceptable levels or they will be liable to sudden failure – or both. This means that you should factor in the cost of batteries when calculating the cost of electric motoring. If your batteries are good for (say) 400 cycles, cost £2000, take you 50 miles on a charge and you cover 10,000 miles a year, your batteries can be expected to last 2 years and therefore cost you £1000 a year or £0.10 pence per mile just for the batteries themselves.

## Temperature

Some batteries operate poorly or have a shorter life if they are used outside their optimum operating temperature range. Discharged lead-acid batteries can freeze and split. The ZEBRA battery (of which more below) has a normal operating temperature over 270 $^o$C (slightly above the maximum temperature of a domestic oven).

## Voltage Sag

When a high current is being drawn from a battery, the voltage will almost always be reduced. This also reduces the power available (recalling that power in watts = volts x amps, so if voltage goes down, power will too).

## Cost

An EV-sized pack of good quality batteries giving you (say) a 100 mile range might costs many thousands of £s. The other problem with batteries is that they tend not to warn you if you are mistreating them. If for example you consistently use 95% depth of discharge, or allow cell imbalance to continue undetected, the first symptoms may become visible only after you have done irreparable damage.

# Battery Principles

Most batteries use *galvanic* action to produce electricity. If you put two dissimilar conductors into a suitable liquid, a voltage is generated between them, and if the two conductors are connected, a current will flow. There is evidence that galvanic cells were in use in what is now Iraq a couple of millennia ago.

You can create a simple galvanic cell very easily. Take a steel pin (like a small nail, an unfolded paperclip or the core of an old pop rivet), plus a similar piece of copper (like the core of the kind of electrical cable used in wiring houses). Acquire a lemon. Push the two pins (one steel, one copper) into the lemon without allowing them to touch. You should be able to measure a voltage between them. The two pins are *electrodes*, and the juice in the lemon is the *electrolyte* (Figure 5-1).

Just how much voltage you will see depends on the materials of which the electrodes are made. A combination of steel and copper will give you up to about 0.8 volt. In theory a lithium or sodium electrode combined with (say) a copper electrode would give you over 3 volts: at least it would do for the split second before it caught fire. Both lithium and sodium are highly reactive and not suitable for use in a battery in metallic form with an aqueous (water-based) electrolyte.

*Figure 5-1 A "Lemon Battery"*

The maximum voltage of material combinations in a battery are governed by their *electrode potential*. Lithium has an electrode potential of -3.04 volts, copper +0.34, so in theory a lithium electrode and a copper electrode used together would give

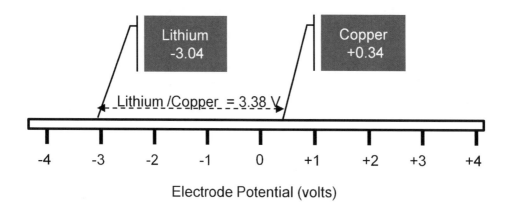

Electrode Potential (volts)

*Figure 5-2 Example of Electrode potential*

you a 3.38 volt battery (Figure 5-2). I should stress again that a lithium-copper battery is purely hypothetical.

Even with a less reactive material than lithium, there are still physical changes occurring in the electrodes of a galvanic cell as power is drawn from it. Sometimes this causes one of the electrodes to be eaten away.

A common example of this is the old-style zinc-carbon battery (also sometimes called a *Leclanche* cell - Figure 5-3) which was

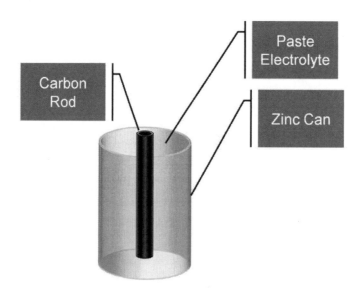

*Figure 5-3 Simplified schematic of Zinc-Carbon (Leclanche) cell*

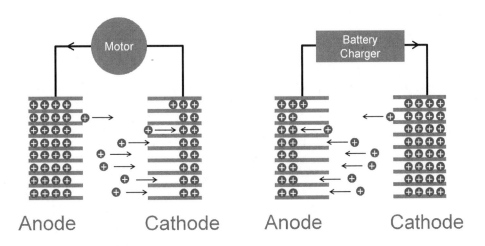

Anode          Cathode          Anode          Cathode

*Figure 5-4 Lithium ion discharge (left) and charge (right)*

the normal household battery before the advent of mass-market manganese-alkali cells. Zinc-carbon cells consisted of a zinc shell or pot with a carbon cathode running up the middle surrounded by a manganese oxide paste. A paste electrolyte was sealed in to the can. As the battery was used, the zinc shell became thinner as the metal was eroded away. Such cells commonly leaked if they were left in equipment after discharge. Lithium sulphur cells are galvanic: the anode is lithium metal and the cathode a mixture of carbon and sulphur.

By contrast, the lithium ion battery is based on the movement of ions (atoms missing an electron or two) between the electrodes, which release and absorb ions in a process known as intercalation (Figure 5-4)

Note that the electrolyte and various other cell components have been omitted from Figure 5-4 for clarity. The small grey disks with the "+" signs represent (positively charged) lithium ions.

There is one other twist to the lithium ion saga. The electrolyte may be a liquid: or it may be a thin flat sheet of a suitable polymer. The latter are the so-called "lithium polymer" batteries. They still operate by intercalation of lithium ions, but the electrolyte is in a different form. Figure 5-5 shows a Kokam lithium polymer cell (left) alongside a Thundersky prismatic lithium ion cell (right). Lithium polymer cells are often (but not always) thin flat pouch style cells.

*Figure 5-5 Left: Lithium polymer. Right: Lithium ion prismatic*

*Photos courtesy of Andrej Pecjak (left) and Jem Freeman (right)*

# Terminology

The "batteries" discussed above should not, strictly speaking, be called "batteries". A battery is a collection of things – in this context, a collection of cells, just as in warfare, a *battery* refers to a collection of guns or missile launchers.

So an ordinary 12 volt car battery is very properly called a battery, because it consists of six cells, each of 2 volts (if you have ever topped up a car battery, you will know that you had six caps or ports where you checked the electrolyte level).

Ordinary mass-market 12 volt car batteries are commonly referred to as *SLI* (Starter, Lighting, Ignition) batteries. For reasons we will discuss later they are not generally suited for use as propulsion batteries in electric vehicles, but are instead optimised to deliver the high current (high power) demanded by starter motors; but only for short periods.

There is a degree of confusion around the terms anode and cathode. *Anode* is usually used to refer to the electrode which is the negative pole of the cell, and *cathode* to the electrode at the positive pole. This is, it must be said, somewhat counter-intuitive.

Cells are divided into two classes. *Primary* cells are use-once-and-throw-away (or recycle). *Secondary* cells can be recharged and used again. So the zinc-carbon cell is an example of a primary cell. The 12 volt automotive SLI battery is made up of secondary cells.

# Specific Energy and Specific Power

Specific Energy and Specific Power are important aspects of a battery. As noted above, the *specific energy* is a measure of the amount of energy (in Joules or in Watt-hours) that the battery is capable of storing per unit of weight (energy density is the volumetric equivalent – energy stored per unit of volume). This is directly linked to the *range* that can be expected from a certain weight or volume of batteries.

The specific *power* is a measure of the power that can be drawn from a battery (measured in Watts, or Joules per second) per unit of weight (volumetric equivalent – power density). This affects *acceleration*

and hill climbing ability.

If energy was orange juice you could think of it like Figure 5-6, On the left is a large juice carton with a big straw. This delivers lots of juice at a high rate. This is analogous to a battery that stores a lot of energy and also delivers high power.

The centre carton in Figure 5-6 is the same size as the previous one but has a much smaller straw. This delivers lots of juice but only at a low rate. This is analogous to a battery that stores a lot of energy but cannot deliver a lot of instantaneous power. The right hand carton is much smaller and

*Figure 5-6 Left High Energy, High Power; Centre High Energy, Low Power; Right low energy. Medium power*

has a medium sized straw. This delivers relatively little juice in total, but is capable of delivering a good flow rate – as long as the juice lasts. This is analogous to a battery that stores little energy but can deliver moderately high power.

Specific energy and energy density tend to be governed by battery chemistry. Lead-Acid batteries have a specific energy around 35 watt hours per kg. This means that if you want 20 kW-hrs from lead-acid batteries (typically good for 40 – 80 miles), the pack would weigh around 570 kgs. Lithium ion batteries have a specific energy around the 100 watt-hour per kg mark or better: so a 20 kW-hr pack might weigh about 200kg. Lithium Sulphur batteries promise 300 watt-hour per kg or better (the 20 kW-hr pack would weigh around 60 or 70 kgs). By contrast, petrol or diesel fuel is good for a couple of thousand watt-hours per kg, so the equivalent of a "20 kW-hr pack" would be around 10Kg of fuel.

Specific power and power density are also governed by cell chemistry, but other factors such as cell design are important too.

# Battery characteristics

There are several battery parameters which it is important to understand if you are going to do a good job of choosing batteries for your vehicle:

## Amp hour rating

The Amp-hour ("Amp-hours", not "Amps per hour") rating is the prime measure of the capacity of the battery. As the term implies it is the number of Amps you can draw multiplied by the length of time that you can draw them before the battery goes flat. In a perfect world, a 20 Amp-hour battery would provide 1 amp for 20 hours or 20 amps for one hour.

Sadly we don't live in a perfect world, so an Amp-hour rating is really only good at a certain discharge rate. A "20 Amp-hour rating at the 5 hour rate" means that the battery is capable of serving up a constant 4 amps for 5 hours; but there are no guarantees about what will happen at higher or lower discharge rate. You might for example find that drawing (say) 15 amps would flatten the battery after an hour (15 Amp-hours) or that that it could supply half an amp for 50 hours (25 Amp-hours).

Murphy's Law means that the true Amp-hour rating of a battery tends to decline as the discharge rate goes up: in other words, if you try to pull a lot of energy out of a battery quickly, you will get less out in total. Furthermore, typical EV discharge rates (30 – 90 minutes) are quite a lot shorter than the industry standard 20 hour rates, so we cannot assume that the Amp-hour rating in the brochure will necessarily apply to a typical EV application.

## Voltage

As we have discussed, cell voltage depends upon chemistry. Furthermore, the percentage difference in voltage between a fully-charged and a fully-discharged cell varies widely. Voltage variations with load current, temperature and even recent charge/discharge history also vary with cell type. Understanding these characteristics is important, and is discussed further both in this chapter and in the one on battery management.

## C rate

The C Rate ("See rate", not "crate") is an important parameter of lithium ion batteries in particular. The maximum C rate of a battery multiplied by its Amp-hour rating gives the maximum current that can be drawn from the battery: so for example a 100 Amp-hour battery with a maximum C

rate of 3 can safely provide 300 Amps, but a 40 Amp hour battery with the same maximum C rate can only provide 120 (40 x 3) amps safely.

A battery may (probably will) have several C ratings: a continuous discharge rate, a continuous charge rate and 1 or more short duration charge and discharge rates. The effect of this may be felt in terms of maximum acceleration and also maximum regeneration. If you fit a big series motor and a controller capable of handling 2000 Amps, but only fit a single string of 40 Amp-hour batteries with a maximum 3C discharge you are going to be seriously disappointed, either by the lacklustre performance, or by the short battery life.

## Peukert number and internal resistance

The C rate is important mostly for lithium ion batteries. For lead acid batteries, the Peukert number is critical. This is a measure of how much the capacity of the battery will drop as the current increases. The lower the Peukert[4] number the better. The number is invariably greater than 1, but

---

[4] The "eu" in "Peukert" is pronounced like "oi" in the English word "oil". I am indebted to Mr Wolfgang Peukert for this information

if it gets as high as 1.3, you have a battery that probably won't do very well in an EV. Suppose that we have four otherwise-identical batteries with different Peukert numbers between 1.0 and 1.3. Suppose that these batteries all have 20 Amp hour capacity at the 20 hour rate and we are drawing 50 Amps from each of them. The variation in time to discharge with varying Peukert numbers is as shown in Figure 5-7.

| Peukert Number | Time to discharge |
|----------------|-------------------|
| 1.0 | 24 minutes |
| 1.1 | 16 minutes |
| 1.2 | 11 minutes |
| 1.3 | 7 minutes |

*Figure 5-7 Discharge time for 20 Ah batteries of different Peukert number at 50 Amp discharge rate*

So a 20 Amp hour battery with a Peukert number of 1.3 would only last 7 minutes at a 50 amp discharge rate. In other words it would in effect be a 6 Amp hour battery under these conditions. This is a pretty serious decline from the 20 Amp hours that you might have been expecting from the manufacturer's spec. sheet. It also represents a serious drop in efficiency. Over 2/3rds of the electrical energy that you put into the battery by charging it would be lost during discharge.

It can be difficult to find Peukert values for specific batteries. However *Reserve Capacity* figures are sometimes published. The Reserve Capacity is a length of time that the battery will last at a higher-than normal output. For example, 12 volt battery Reserve Capacity is frequently measured at 25 Amps – so a 100 minute reserve capacity represents 100 minutes at 25 Amps.

# Batteries for EVs

At the time of writing there are just two battery chemistries in common use in amateur conversions:

- Lead Acid
- Lithium Ion

Another chemistry which is just becoming available commercially, and which could be a game-changer is:

- Lithium Sulphur

In addition there are three more chemistries that are or have been used by some commercial builders:

- *Nickel Cadmium* – NiCad (Now obsolete). Was used in Citroen

Berlingo Electrique and Saxo Electrique)

- *Nickel Metal-Hydride* - NiMh (Used in Toyota and Honda hybrids, and the Toyota RAV4 EV)
- *Molten Salt* (e.g. ZEBRA used in some Modec vans and a 1997 prototype electric Mercedes A class)

These latter are worth being aware of. There is a lot of battery research going on so new technologies (or new adaptations of existing technologies) may pop up at any time.

# Lead Acid

All lead-acid batteries use the same basic electrode materials – a *lead* anode and a *lead dioxide* cathode. This combination produces just over 2 volts per cell. In most lead-acid batteries, the electrodes are not pins or rods but flat plates. The electrolyte is sulphuric acid. There are three classes of lead acid batteries, and two (or maybe three) variants in each class.

The three classes of Lead Acid batteries are

- Flooded Lead Acid (sometimes "FLA" or just "Flooded")
- Absorbed (or Absorbent) Glass Mat – AGM
- Gel

AGM and Gel batteries are sometimes referred to as "Sealed", "Valve Regulated Lead Acid" (VRLA) or "Starved Electrolyte" batteries. In addition, lead acid batteries are characterised by their application: SLI, Traction/Deep Cycle or Leisure:

## SLI

(Starter, Lighting, Ignition) batteries are designed to deliver lots of power but only for a short time. They are intended to be discharged just a few percent in normal operation. Regular deep discharging destroys them very quickly

## Traction

Or *Deep Cycle* batteries on the other hand are designed for regular deep discharge. Their specific power is lower than an equivalent SLI battery, but their deep-discharge cycle life is a lot better.

## Leisure

There is sometimes a third category: the *Leisure* battery. This is intended to be used in boats, caravans etc and is a compromise: better cycle life than the SLI battery, and better specific power than the traction battery.

There are therefore in theory 9 types of Lead Acid battery as shown in Figure 5-8.

|  | SLI | Leisure | Traction |
|---|---|---|---|
| **Flooded** | X | X | X |
| **AGM** | X | X | X |
| **Gel** | X | X | X |

*Figure 5-8 Taxonomy of Lead-Acid batteries*

## Flooded Lead Acid

The standard SLI (car) battery that most people know is usually a "Flooded Lead Acid" (FLA) battery. Each cell consists of a plastic box containing flat plate electrodes immersed in sulphuric acid. The basic principle hasn't changed since the late 1800s.

Each cell will have many plates interleaved with separators (anode-separator-cathode-separator-anode-separator...). The trick is to maximise the plate area. This minimises internal resistance and maximises output. Unfortunately the thinner the plates, the lower the tolerance of the battery to repeated discharge. SLI batteries have thin plates, so you can pack more plates into a given volume and as a result they can sustain high currents. Traction batteries have fewer, thicker plates, so last longer (Figure 5-9).

## Absorbed Glass Mat (AGM)

AGM batteries do not contain free liquid electrolyte: instead they have separators made of a form of glass fibre which holds

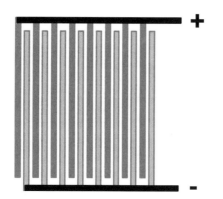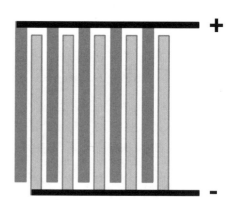

*Figure 5-9 Plate Layout: SLI (Left) Traction (Right)*

the electrolyte. These batteries have several advantages over flooded batteries. They:

- will not give off any gas or fumes if you treat them right.
- never need watering
- tend to have a higher specific power than flooded cells
- are said to be much more shock and vibration resistant
- can be mounted in almost any orientation

They also have a reputation for low Peukert number (meaning that they retain a reasonable capacity at high discharge rates as described above). It is however important not to overcharge them. If overcharged they may start to vent. Regular venting will kill the batteries. This more or less mandates a good quality battery charger. AGM batteries are also the most expensive of the three lead acid cell types (Flooded, AGM, Gel).

## Gel Batteries

Gel batteries are a kind of half-way house between Flooded and AGM. The acid electrolyte has chemicals added to it which turn it into a jelly. Like AGM, Gel batteries should not leak or vent fumes in normal use and should never need watering. It should also be possible to mount them in almost any orientation.

# Some examples of Lead-Acid batteries

### OPTIMA

OPTIMA (http://www.optimabatteries.com/) sell a range of batteries, all AGM, with an unusual design: the plates in each cell are a single long strip wound into a roll like an old-fashioned elastic bandage and inserted into a cylindrical cavity in the casing (see Figure 5-10).

The batteries are colour coded by application: the "Red Tops" are SLI batteries. The "Blue Tops" have a slightly different terminal arrangement for Marine applications and are available as SLI or deep-cycle. The Yellow Tops are deep-cycle batteries. A light grey case denotes deep cycle, a dark grey case SLI.

The OPTIMA D51 has a 38 Amp-hour capacity at the 20 hour rate, and a reserve capacity of 66 minutes. Peukert is about 1.12, so it could deliver 50 amps for about 30 minutes (equivalent to 25 Amp hours at that rate). It weighs just under 12 kg.

*Figure 5-10 OPTIMA Batteries. Note extra terminals on one of the batteries (a "Blue Top").*

*Photo courtesy of OPTIMA Batteries (http://www.optimabatteries.com)*

## Trojan

Trojan Battery has more than 85 years of battering engineering and manufacturing experience. The T-105 model (Figure 5-11) is the company's flagship deep-cycle flooded lead acid battery. It is 6 volts and is fairly heavy, so 24 of them weighing close to 700kg would be needed to make a typical 144 volt pack. However, it does have a high amp hour rating (225 at the 20 hour

rate), so it should not be compared to the other batteries featured here which have lower amp hour ratings. Its Peukert is 1.16, and because of its large capacity it can provide 50 amps for about 3½ hours.

## Haze

The Haze HZY-EV12-33 (Figure 5-12) is a 12 volt gel battery with a 29 Amp hour capacity at the 20 hour rate, falling to 18

*Figure 5-11 Trojan T-105. Photo courtesy of Trojan Battery (http://www.trojanbattery.com)*

*Figure 5-12 Haze Gel Battery. Photo courtesy of Haze Batteries Europe Ltd (http://www.hazebattery.com/)*

Amp hours at the 1 hour rate. This is equivalent to a Peukert value of about 1.12. It could deliver 50 amps for about 17 minutes (equivalent to 14 Amp hours at that rate). It weighs 10.4 kg.

## Enersys Odyssey

Anecdotally, Enersys batteries have a good reputation in the EV world. The PC925 (Figure 5-13) is another AGM battery. It has a capacity of 28 Amp hours at the 20 hour rate and a low (= good") Peukert value of about 1.09. At 50 amps it would last about 24 minutes for an equivalent of 20 Amp hours. It weighs just under 12 kg. The PC1750 (Figure 5-14) is a larger Enersys battery. It has a capacity of 74 Amp hours at the 20 hour rate. The Peukert exponent is about 1.14. At 50 amps it would last about an hour. It weighs 26.3 kgs.

## CSB

The CSB 12220-X2 is an example of a battery with a relatively high Peukert number. It is a 20 Amp-hour battery (measured at the 20 hour rate) weighing about 6.7 kg. Peukert, though, is around 1.2 so it would only maintain a 50 Amp discharge for 11 minutes or so; equivalent to about 9½ Amp hours, or less than half what you might expect from the headline Amp-hour rating.

*Figure 5-13 Enersys Odyssey PC925. Photo courtesy of Enersys (http://www.odysseyfactory.com)*

*Figure 5-14 Enersys Odyssey PC1750s in a bike frame. Photo courtesy of Jason Siegel*

# Lithium Ion Batteries

Lithium ion secondary cells have huge theoretical advantages over lead acid. They have three or four times the specific energy of lead acid and can in some cases deliver good specific power and long cycle life. On the other hand, lithium ion technology is still developing. The chemistry that was dominant as recently as 2 – 3 years ago is now almost unobtainable. It is also an area where conspiracy theorists can have a field day: until quite recently, most US and European suppliers of lithium ion cells would not sell batteries for use by private individuals; they would only talk to vehicle manufacturers. This situation is changing however.

There are other issue with lithium ion batteries:

- They need to be looked after very carefully. More on this later.
- A large capacity lithium ion battery pack can easily cost tens of thousands of pounds from some suppliers.
- Neither the technology nor the market is mature and independent verification of things like cycle life does not, in many cases, exist.

This means that going with lithium ion carries great promise (100 mile + range is reported by some of those using them) but it is also risky. One respected author[5] recommends giving lithium ion a miss for the time being. Others with experience in the field consider them to be the only battery type for a usable electric vehicle rather than for experimental machines. We will look at them in a little more detail and it is then up to you to make a choice based upon the depth of your wallet and your range aspirations. First a little more theory.

## Elements used in Lithium Ion

Lithium ion chemistries are often labelled using the standard chemical symbols for the elements used in the cathode. For example "Lithium-Cobalt" cells (denoting cells with a cathode made of lithium cobalt) are often shortened to **LiCo**. To add further confusion, this battery chemistry is properly referred to as "lithium cobalt oxide" or **$LiCoO_2$**.

The main chemical symbols to be aware of are:

**Fe** – Iron (NB – *I*r*on* not *Ion*)

---

[5] Mike Boxwell: 'Owning an Electric Car' 2010 edition

**H** - Hydrogen

**Li** - Lithium

**Mn** – Manganese

**Ni** – Nickel

**O** – Oxygen

**$O_2$** – Usually denotes an oxide in this context

**P** – Phosphorus

**$PO_4$** – Phosphate

**S** – Sulphur (or Sulfur in American English as adopted by the International Union of Pure and Applied Chemistry - IUPAC)

**Ti** - Titanium

**Y** – Yttrium

Subscript numbers indicate the number of atoms in a molecule: so for example **$H_2O$** (water) is made up of molecules containing 2 Hydrogen atoms (**$H_2$**) and one Oxygen atom (**O).** Manganese Dioxide is **$MnO_2$** (one Manganese atom, two oxygen atoms per molecule)

One of the main Lithium-Ion cell chemistry in use today is Lithium Iron Phosphate (**$LiFePO_4$** – sometimes shortened further to "LFP"). There are a few further chemistries that you may encounter:

- Lithium Cobalt Oxide (**$LiCoO_2$** – dominant until recently)
- Lithium Manganese Oxide

(**$LiMn_2O_4$**)
- Lithium Nickel Oxide (**$LiNiO_2$**)
- Lithium Nickel Manganese Cobalt Oxide (**$LiNiMnO_2$** – or "NMC")

All of these refer to the material used for the *cathode*. Normally the anode of a lithium ion cell is graphite, but some batteries use other materials. If you encounter a "lithium titanate" cell, this is a reference to the *anode* material which is (confusingly) also a lithium compound – lithium titanate or **$Li_2TiO_3$**.

Note that the dominance of graphite as an anode material in lithium ion cells may not be permanent. It is far from the best material electrochemically, but better alternatives (such as Tin – symbol **Sn** - and Silicon - **Si**) degrade very quickly. Changes in volume on each cycle rapidly pulverise the anode. One possible approach is to encapsulate these materials in carbon nanotubes.[6]

Lithium-air cells are also the subject of research. In addition there are all manner of doping and tweaking additives which can affect the performance and characteristics

---

[6] For more detail see for example "Tin-based materials as advanced anode materials for lithium ion batteries: a review" by Ali Reza Kamali and Derek J. Fray

of the battery. Note that different lithium ion chemistries have different specific energies and cell voltages.

## Lithium Battery Performance

There are several factors of importance with lithium ion batteries:

**Voltage** - nominal voltages vary from about 3.3 volts per cell to 3.8 volts per cell.

**Price** – expressed in £ or $ per watt-hour. This can vary very widely between suppliers.

**Failure rate** – anecdotal evidence suggests that a small proportion of any given batch of lithium ion cells may either be dead on arrival or fail after a short period of use.

**Specific Power, Specific Energy and C rating** – as discussed earlier. There is quite a lot of variance here between different lithium ion cell types.

**Cycle life** – again there is a lot of variance from a few hundred cycles to several thousand; but as noted above independent evaluation is hard to come by.

## Lithium Cobalt Oxide (LiCoO$_2$)

Lithium Cobalt Oxide cells were dominant until just a few years ago. They have rather fallen out of favour however because there were a number of cases of these batteries overheating or exploding in consumer electronics such as laptops. Compared with other lithium ion chemistries they have:

- High specific energy
- Lower C rating (i.e. low maximum discharge rate and hence poor specific power)
- Shorter cycle life

In summary they would take you a long way on a single charge, but acceleration and hill climbing ability would be mediocre. These cells had a reputation for catching fire if abused, and they would normally need replacing every year or two.

## Lithium Iron Phosphate

LiFePO$_4$ or "LFP" is dominant for new lithium ion batteries shipping today for amateur use. Compared with lithium cobalt oxide, lithium iron phosphate cells tend to have a slightly lower specific energy (although still far better than most non-lithium chemistries). Otherwise they are a winner all round: safer, higher C rating and longer life. Proprietary cells shipping today often appear to be available in two forms: optimised for specific energy or optimised for specific power.

## Other Lithium ion Chemistries

I am not going to say very much about other lithium ion chemistries, mostly because at the time of writing there are few sources of such lithium ion cells or batteries using them. Lithium manganese oxide ($LiMn_2O_4$) appears to be about mid way between lithium cobalt oxide ($LiCoO_2$) and lithium iron phosphate ($LiFePO_4$) in safety, specific energy and C rating.

You may also encounter Lithium Nickel Manganese Cobalt Oxide ($LiNiMnCoO_2$) [Kokam and Panasonic] and Lithium Cobalt Nickel Aluminium Oxide ($LiCoNiAlO_2$) [Panasonic].

## Lithium ion Battery Packs

As described earlier, lithium ion cells may have a liquid electrolyte, or they may be polymer cells. The latter are typically (but not necessarily) a thin bag or pouch. The former are typically available as either cylindrical cells (not unlike an overgrown domestic AA or D cell) or as "prismatics".

The cylindrical cells are typically in the 2 – 8 Amp hour range, whereas prismatic cells are available in capacities of 200 Amp hour or even higher. A 100 Amp hour prismatic cell is about the size of a small but thick book or a small box of breakfast cereal. One popular 100 Amp-hour cell is 220mm (8.66") by 179mm (7.04") by 62mm (2.44") and weighs about 3.5 Kg

Making up a battery pack with Prismatic cells is mostly a matter of constructing a

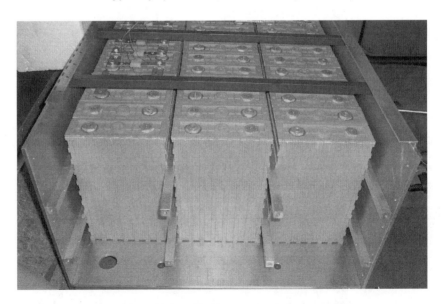

*Figure 5-15 A battery pack of prismatic cells under construction.*
*Photo courtesy of Greg Sievert*

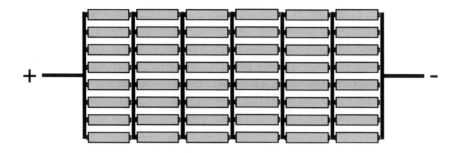

*Figure 5-16 Schematic of cylindrical cell battery pack (6 clusters of 8 cells each)*

suitable battery box (or boxes) and chaining the cells together using suitable straps or cables (Figure 5-15).

Some types of Lithium cell are commonly fitted with rigid end plates to prevent swelling of the cells. Be guided by the manufacturer's recommendations.

Using the cylindrical cells is a little more complicated. You need to assemble clusters of cells in parallel to get the Amp-hour rating that you need, and then assemble the clusters in series until you have enough voltage (Figure 5-16).

Alternatively you might be able to assemble multiple high voltage strings in parallel (Figure 5-17).

Some of these cylindrical cells have threaded terminals. Others have tabs that can be welded and with yet others you will need to flash-weld tabs or bus bars to the battery terminals. This sounds fiddly, but does have a couple of advantages. Firstly, if a cell fails the effect is less severe than with a single string of big prismatic cells. If you have a battery pack made up of prismatic cells in series, you have a chain that is only as strong as its weakest link.

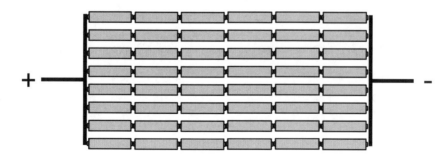

*Figure 5-17 An alternative topology; multiple strings*

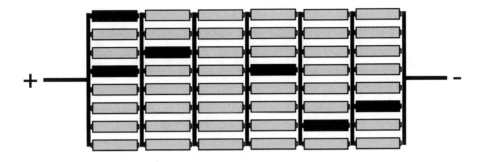

*Figure 5-18 Battery pack with random failures*

Suppose for example you have such a battery pack and that ordinarily it gives you a 100 mile range. Now suppose that one of your cells partially fails and reaches minimum voltage after just 30 miles. You have two choices when the weakest cell reaches its minimum voltage: stop and recharge, or carry on and almost certainly destroy the bad cell altogether – and (with some chemistries) risk wider pack damage.

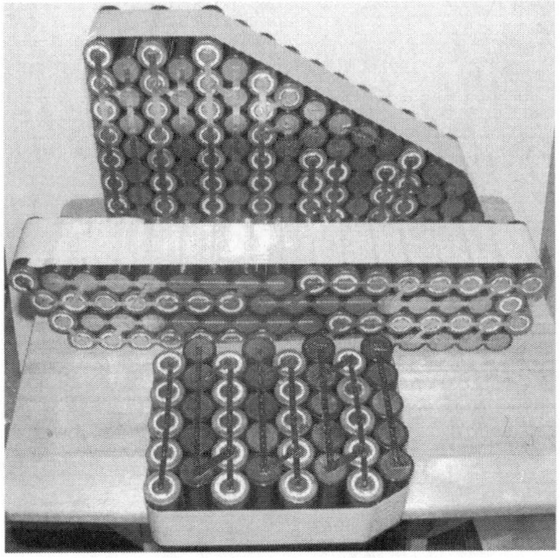

With a large number of smaller cells, any degradation is more graceful. Unless you are unfortunate, a few random cell failures are only going to cut down your range by a few percent (Figure 5-18)

The second advantage of using clusters of smaller cells is that you have more flexibility in use of space. Figure 5-19 for example is a pack made up of cylindrical cells intended to fit into spaces in the frame of an electric bike:

The disadvantage with this arrangement is what can go wrong with it. Using small cylindrical cells in a full size EV implies many hundreds if not thousands of high-current connections. Every one is a potential point source of overheating.

*Figure 5-19 Battery Pack for use in an electric bike. Photo courtesy of Falcon EV (http://www.falconev.com)*

Common Cylindrical Cell Sizes

| Designation | Nominal Diameter (mm) | Nominal Length (mm) | Approximately equivalent to |
|---|---|---|---|
| 10440 | 10 | 44 | AAA |
| 14500 | 14 | 50 | AA |
| 17500 | 17 | 50 | A |
| 26460 | 26 | 46 | C |
| 26650 | 22 | 65 | |
| 32650 | 32 | 65 | |
| 33580 | 33 | 58 | D |
| 602030 | 60 | 203 | |

## Cylindrical Cell Sizes

Cylindrical cells are usually designated by diameter and length (albeit inconsistently). For example a "26650" cell is about 26 mm in diameter and 65 mm long (or roughly 1" diameter and 2½" long). Some common sizes are listed above. Note that the diameter and length may not be exact. There are differences between cells from different manufacturers. Also, the "length" may be that of the cylindrical part of the cell and may exclude terminals on the ends. Lastly (and confusingly) a "38120" may be 120 mm rather than 12mm long.

# Examples of Lithium ion cells

Here are some examples of Lithium ion cells. Note that these are manufacturer's data and in some cases it's prudent to take some figures (particularly C rating and cycle life) with a large pinch of salt.

## A123 ANR26650 M1

A123 are a US company that manufactures lithium ion cells currently regarded as amongst the best available. Probably their best known cell at the time of writing is the ANR26650M1A (Figure 5-20).

This cell weighs 70 grams, delivers 2.3 Amp hours at 3.3 volts and astonishingly can sustain 70 Amps continuous discharge rate, or 120 Amps for 10 seconds. The latter is a C rate of over 50 (anything over 3 would until quite recently have been regarded as good). A123 claim a cycle life

*Figure 5-20 A123 ANR26650M1A cells. Photo courtesy of Albert Van Dalen*

in excess of 1000. Plus this cell has a wide usable temperature range: it should be usable anywhere from the arctic to the tropics. A123 have recently produced a pouch-style cell which looks interesting (albeit fairly expensive).

## Thundersky LFP160AHA

Thundersky (aka "Winston") have been one of the most popular makers of lithium batteries amongst amateur builders and converters, although at the time of writing there was uncertainty over their future. They are based in China and specialise in large prismatic cells. Their popularity is probably explained by a combination of price and availability. A few years ago, their cells were mostly lithium cobalt oxide ($LiCoO_2$), with a relatively poor C rating (you couldn't suck power out of them too quickly) and like most cells based on this

chemistry there were questions over safety. Nominal life was a few hundred cycles.

Their current generation of cells are lithium iron phosphate ($LiFePO_4$ – see Figure 5-21). Like all $LiFePO_4$ cells they have a nominal 3.2 volt output, should be fairly

*Figure 5-21 Thundersky LFP160AHA 160 Amp-hour Cell (209×280×65 mm, 5.6 kg). Photo courtesy of Jem Freeman*

safe i.e. less likely than $LiCoO_2$ to explode if punctured. They do have a slightly lower specific energy than other lithium ion chemistries, albeit still much better than for most non-lithium chemistries. This translates into a somewhat shorter range for a given battery pack weight than an equivalent lithium cobalt oxide pack.

This cell has a maximum continuous C rating of 3 and a peak of 20. For a 160 Amp-hour cell, this translates into 480 Amps continuous and 3200 Amps peak (it is not clear for how long this peak can be maintained). Like most manufacturers of $LiFePO_4$ cells Thundersky claim a cycle life in excess of 2000. If this was achieved, the cells could be charged every weekday for almost 10 years.

## Headway

Headway are another Asian manufacturer. Like Thundersky, their current product range is dominated by Lithium Iron Phosphate ($LiFePO_4$) cells i.e. safer, longer life, higher specific power but slightly lower specific energy than Lithium Cobalt Oxide ($LiCoO_2$).

Unlike Thundersky however, Headway produce cylindrical cells (Figure 5-22). At the time of writing they offer a couple cell sizes. Their 38120 cells are interesting. They are all, as the name implies nominally

*Figure 5-22 Headway 38120 Cell. Photo courtesy Tony Castley*

38 mm (about 1½ inches) in diameter and 120 mm (just under 5 inches) long. All have a nominal voltage of 3.2 and all weigh about 0.3 kg. The 38120S is optimised for specific energy (10 Amp-hours) with a maximum continuous C rating of 15 (150 Amps). The 38120P and 38120HP are slanted the other way: capacity is slightly lower at 8 Amp-hours but maximum continuous C rating claimed is 25 (equivalent to 200 Amps, given the slightly reduced capacity). All other things being equal, the 38120P and HP would give you more acceleration away from the lights but 20% less range.

Unlike the A123 cells discussed earlier, Headway cells are available and affordable, albeit rather less capable than their A123 equivalents,

## Sky Energy/CALB

China Aviation Lithium Battery (CALB – formerly Sky Energy) cells (Figure 5-23) are in many respects similar to their Thundersky equivalents: affordable and available lithium iron phosphate prismatics. They even look similar apart from colour. Nominally the Sky Energy/CALB cells have a slightly higher continuous C rating than the Thunderskys (4 rather than 3)

*Figure 5-23 Sky Energy/CALB 130 Amp-hour cell. Photo courtesy of Greg Sievert*

## Kokam 100216216H

Kokam Cells are lithium polymer type – thin flat cells. The 100216216H (Figure 5-24) is a 3.7 volt 40 Amp-hour cell weighing just over 1 kg. The body of the cell (ignoring the tabs) is about 220 mm high and 215 mm wide – but only about 11mm thick.

*Figure 5-24 Kokam 100216216H 40 Amp-hour cell. Photo courtesy of Victor Tikhonov (http://www.metricmind.com)*

The cathode material used in most of the latest Kokam cells is Lithium Nickel Manganese Cobalt Oxide ($LiNiMnNiCoO_2$ – often shortened to "NMC" fortunately). Anecdotally, this chemistry appears to be safer than the older Lithium Cobalt.

## Battery Packs

This is a different approach. All the foregoing examples are single cells with a nominal voltage around 3.2 – 3.7. Several manufacturers offer complete battery packs containing many cells and (usually) a Battery Management System (BMS).

Packs like these would appear to have advantages for use in EVs and HEVs but to date do not appear to have achieved great popularity. There are also horror stories circulating about poorly-constructed packs with dubious BMS systems: so it is definitely a case of *caveat emptor*.

# Lithium Sulphur

At the time of writing, Lithium Sulphur (Li-S) cells were only just coming onto the market in sizes, and with a cycle life, to make them practicable for EVs. Information about their performance in EVs is therefore in short supply.

Despite the similarity in naming they are quite different from lithium ion cells. You will recall that lithium ion cells operate by intercalation. Lithium sulphur cells on the other hand are galvanic like lead-acid cells (and like the "lemon battery" illustrated in Figure 5-1). The anode is metallic lithium and the active ingredient of the cathode is sulphur. Lithium reacts violently with water so the electrolyte is an organic solvent. The exciting thing about these cells is their potential specific energy: at least three times that of Lithium Iron Phosphate ($LiFePO_4$). Li-S cells should also turn out to be cheaper in quantity and early indications are that they are safer.

# Nickel-based Chemistries

Until recently, most volume manufacturers of hybrid vehicles used either Nickel Cadmium or Nickel Metal Hydride cells. These have a better specific energy than Lead Acid, but far lower specific energy than lithium ion or lithium sulphur. Their characteristics were well known however and they were generally safer, longer lasting and more reliable than earlier generations of lithium-ion cells.

## Nickel Cadmium

Flooded nickel cadmium (NiCd or "Nicad") cells were used in the Citroen Berlingo Electrique and the Citroen Saxo Electrique. These cells are said to be very tolerant of abuse and have long cycle life but are now obsolescent. Nickel cadmium cells are in fact banned in Europe for most purposes because cadmium is regarded as a hazardous material from a health-and-safety viewpoint. Nicads are generally robust but do have one or two potential problems. The internal resistance of a nickel cadmium cell drops as it approaches full charge and also drops as temperatures rise. With a powerful but relatively crude charging system, this can lead to thermal runaway, fire and/or explosion. When

Nickel cadmium batteries were first introduced into aircraft there was some concern about this. For example in August 1971 a number of people on a Viscount aircraft had a fortunate escape when a Nickel Cadmium battery suffered thermal runaway as a result of an accidental short circuit. The heat from the battery destroyed some flight control pushrods: fortunately the aircraft had just landed. A few minutes earlier and all on board might have been killed.

The chemistry of Nickel Cadmium batteries:

**Cathode**: Nickel Oxy-Hydroxide (NiOOH)

**Anode**: Cadmium (Cd)

**Electrolyte**: (typically) Potassium Hydroxide (KOH). This is an alkali rather than an acid.

## Nickel Metal-Hydride

In a world free of patent encumbrances and commercial interest groups, it is possible that nickel metal-hydride might have been the dominant battery chemistry in EVs and HEVs. These batteries have lower specific energy and lower specific power than equivalent lithium chemistry cells but they

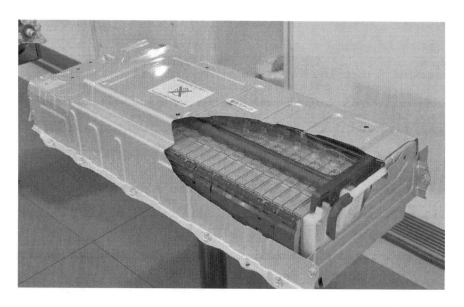

*Figure 5-25 Nickel Metal Hydride Battery Pack from a Toyota Prius. Photo courtesy of Randy Hsieh (http://www.flickr.com/photos/palto243)*

appear to be safe, reliable, easy to manage, long-lived and (above all) much cheaper.

Nickel metal-hydride batteries have dominated in volume production EVs and HEVs. They are or were used in the following cars amongst others:

- Toyota RAV4 EV
- Toyota Prius (Figure 5-25)
- Honda Civic Hybrid
- Honda Insight
- Ford Ranger EV (later versions)
- Ford Escape hybrid
- General Motors EV1 (1999 and on)
- Various Lexus hybrids

The chemistry is similar to the nickel cadmium but without the cadmium. Cycle life of nickel metal-hydride is better than lead acid but not as good as nickel cadmium.

**Cathode**: Nickel Oxy-Hydroxide (NiOOH)

**Anode**: A metal alloy capable of storing hydrogen

**Electrolyte**: (typically) Potassium Hydroxide (KOH)

The use of multiple small batteries in parallel as described above for lithium ion is problematic with nickel metal hydride: cell voltages drops slightly as the cell reaches full charge, so if two cells are charged in parallel, a fully charged cell will draw more current than its undercharged neighbour.

Considerable controversy has however surrounded large format nickel metal hydride cells. Two things are clear:

Firstly, until recently, nickel metal hydride patents were controlled by a large multinational oil company. Secondly, large format nickel metal-hydride batteries have been virtually unavailable for some years.

There have been widespread allegations that these two facts are connected and that the reason was a desire to constrain the growth of electric vehicles. As of early 2010, there have been reports [7] that suggest that the acquisition of Nickel metal-hydride patents by SB LiMotive (a Bosch-Samsung joint venture) might lead to greater availability of these batteries.

It might not however. The main business of SB LiMotive is the production of lithium ion batteries. It is possible that they might view nickel metal-hydride as a competitor and continue to block its use in EVs. Watch this space.

---

[7] See for example

http://www.sae.org/mags/AEI/7552

# Molten Salt Batteries

Molten Salt batteries, as their name implies, use a salt in liquid form as the electrolyte: and so (normally) operate at high temperatures.

The pre-eminent molten-salt battery available today (and almost the only one available commercially) is the ZEBRA battery (Figure 5-26). This has a cathode of liquid sodium, and a nickel chloride anode. The electrolyte is Sodium Aluminium Chloride ($NaAlCl_4$). In many ways it is very attractive package; it has good specific power, fairly good specific energy (better than nickel metal-hydride but not as good as lithium ion), and if manufactured in quantity would be fairly cheap. The design has been around since the eighties, the active materials are cheap and plentiful, and the cycle life is both proven and good. These batteries can be recharged quite rapidly. They appear to be both safe and robust (neither overcharging nor over-discharging does them much harm).

Sounds great, but ZEBRA batteries have one big drawback. Their normal operating temperature is around 300 degrees Centigrade. This isn't hazardous – the outer surface of the battery is just warm to the touch – but it adds complexity to the control systems and means that the battery normally consumes some power when not in use just to stay at normal operating temperature. The power consumption to keep the battery warm enough to be available for immediate use is modest – about the power required by a domestic incandescent light bulb. The battery can be allowed to cool and freeze, but reheating it takes 12 – 24 hours.

As mentioned above, the Zebra battery was used in some Modec vans. It has also been used in some conversions.

*Figure 5-26 ZEBRA battery.*
*Photo courtesy of John Honniball*

# Batteries - in Conclusion

If you are building an EV/HEV, or converting a car to electric power, perhaps the biggest and most complex decision you will have to make is the choice of battery for your project. If you choose one of the more sophisticated chemistries, the batteries will probably represent the largest single element in the cost of the project. The batteries will undoubtedly be the heaviest and bulkiest component regardless of chemistry.

If you choose poorly, no amount of sophistication in the controller, the motor or other components will make up for a battery that cannot store enough energy and/or cannot deliver it fast enough to maintain an adequate speed up a long incline.

It is also the area where you will need to do the most extensive research. For each type of motor, there are only a few manufacturers and models, but the range of potential battery suppliers is vast. There are thousands of varieties of lead acid batteries available. During the research for this book around 40 different brands of lithium ion batteries were identified: and almost all offered a range of sizes and formats.

We have now looked at the three core elements of an EV drivetrain: motors, controllers and batteries. In the next chapter we will look at the thorny topic of managing your batteries: controlling charge and discharge, monitoring their health and optimising their useful life.

# 6: Battery Management

It is arguable that battery management practices, more than any other factor, determine the safety, practicality and total cost of ownership of an EV or HEV.

A typical lithium ion battery pack for an EV costs thousands of £s. It is easy to destroy it in a single over-charging or over-discharging incident. If that happens after, say, 100 charging cycles, rather than the 2000 cycle design life of the individual cells, the economics of the project are just untenable: it would have been cheaper to convert the vehicle to run on single malt whiskey. Worse, if you are unfortunate, that single overcharging incident could burn out your vehicle and maybe even burn down your house.

Moreover, battery management is an area where there is, for some cell types, little hard research evidence, some anecdotal evidence and much heated debate. We will nevertheless attempt to pick our way through the minefield.

Battery management comes down to three things:

- How you manage the charging process
- How you manage the discharging process
- How you manage long periods of inactivity

So as part of this chapter we will discuss chargers (as physical bits of kit). If your 12 volt SLI battery won't start your IC engine, you can acquire a charger for a few pounds almost anywhere, connect up and switch on and your battery will charge up. Charging the battery of an EV is sadly not so straightforward. The wrong charger, or the right charger wrongly used will kill your battery pack.

In this chapter we will build up slowly

- Terminology
- General principles
- Managing single cells

- Managing entire battery packs

- Pack balancing/equalisation

- Automated battery management systems (BMSs)

- Estimating remaining charge ("fuel gauges")

- Fast charging

- Real world chargers

We will focus largely on lead acid and lithium ion cells as these are by far the most common types in EVs being built today. We'll also touch on what little is known of the latest lithium sulphur cells. Many of the same principles carry over to other cell chemistries.

First of all, though a word of caution. There is real-world in-service experience with lead acid propulsion batteries in cars going back to the end of the 19th century. Experience of Nickel metal-hydride (NiMh) batteries in various mass-produced Citroen, Peugeot and Toyota EVs goes back to the mid '90s.

Lithium ion is different. There are billions of small format lithium ion cells in laptops, mobile phones, and even model aircraft all over the world, but the use of long strings of big lithium ion cells in electric vehicles is a recent phenomenon. With few exceptions (such as Modec vans, Tesla and some Chinese manufacturers), the use of these battery packs in EVs has been pioneered by individual enthusiasts. That situation is changing with a number of volume manufacturers beginning to offer EVs powered by lithium ion cells: but most of the world's experience of large format lithium ion batteries in EVs has been accumulated since about 2005 and even by 2010 didn't amount to much. Furthermore, lithium ion cells are not all the same. As we discussed in the previous chapter, lithium cobalt and lithium iron phosphate cells may look similar but they operate at different voltages and behave differently when mistreated (lithium cobalt cells are less stable than lithium iron phosphate).

# Terminology

Firstly then, a reminder of terminology. You will recall from the chapter on batteries that the fundamental unit in a battery is the *cell*, each cell having an anode, a cathode and some form of electrolyte. The nominal cell voltage is a factor of the chemistry: about 2 volts for a lead-acid cell, 3.2 volts for a lithium iron phosphate cell and maybe 1.2

volts for nickel metal hydride.

A *battery* is as a collection of cells usually connected in series, but sometimes a combination of series and parallel. A normal 12 volt automotive SLI (Starter, Lighting, Ignition) battery is 6 individual lead acid cells in series; these short strings packaged and sold as a single unit are sometimes referred to as *monoblocks*. Many EVs are powered by strings of monoblocks (e.g. 10 x 12 volt batteries for a nominal 120 volts). Others are powered by series strings of individual cells (e.g. 50 x 3.3 volts for a nominal 165 volts). Herein I will refer to a vehicle-scale set of monoblocks or cells as a *battery pack* or a *traction pack*.

# General Principles

The problem with managing battery packs is that each cell potentially needs treating differently, but for practical purpose a pack must be charged and discharged as a unit using the same current/time profile. The problem gets worse as the number of cells in the pack grows. As if this wasn't enough to contend with, there are a number of complicating factors.

## Complicating factor no 1: voltage and state of charge

Anyone who has messed around with the 12 volt electrics in a conventional car will have an intuitive understanding that open circuit voltage (voltage when there is no current flowing) varies with state of charge (SOC). Indeed the rested open circuit voltage (OCV) of a lead acid battery varies

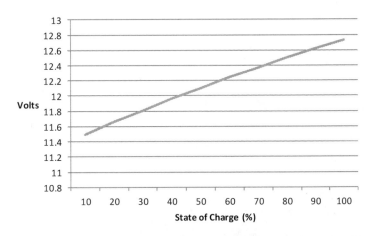

*Figure 6-1 Voltage versus State of Charge for a Rested Lead Acid Cell*

roughly linearly with state of charge ("rested" = "hasn't been charged or discharged recently"). See for example Figure 6-1 which is based on data from a Trojan battery.

Not all cell chemistries behave like this however. Nickel based cells (nickel-cadmium, nickel-metal hydride) exhibit a small voltage *drop* near 100% state of charge (Figure 6-2). This is seriously confusing for anyone used to lead acid batteries. Note that Figure 6-1 is not directly comparable with Figure 6-2, because Figure 6-1 refers to a battery, and Figure 6-2 to a single cell.

With NiMh cells, once the cell voltage starts to drop at the end of the charging cycle, cell temperature and cell internal pressure start to rise rapidly.

Lithium ion cells typically display an "S" shaped relationship between state of charge and voltage. The exact form of this and the precise voltages vary with cell construction and chemistry, but broadly there is a pronounced knee near minimum state of charge and another near maximum state of charge (See Figure 6-3).

Note that voltage versus state-of-charge curves are usually a similar shape, but not exactly the same values, for charge and

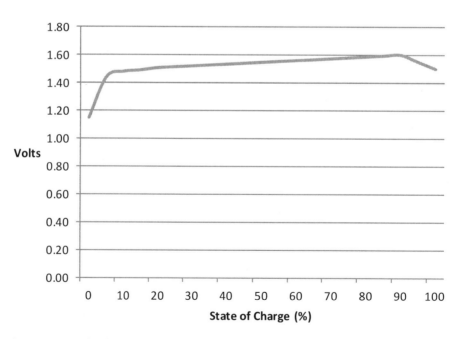

*Figure 6-2 Typical Voltage versus State of Charge for Nickel Metal Hydride Cell*

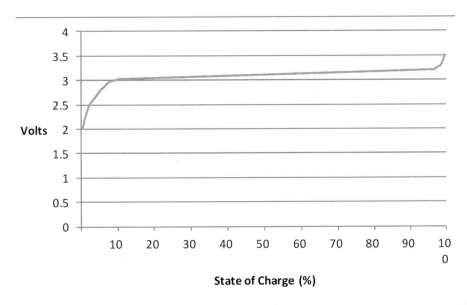

*Figure 6-3 Typical Voltage versus State of Charge for Lithium Ion Cell*

discharge.

## Complicating factor no 2: Internal Resistance

All cells exhibit some degree of internal resistance. From the outside, this behaves as if you had a perfect battery connected in series with a low resistance. If the internal resistance is (say) 0.001 ohm then the terminal voltage of the cell will drop 0.1 volts at 100 amps (Volts = Amps x Resistance: 0.001Ω x 100A = 0.1 volts: (Figure 6-4).

Cells are not of course fitted with discrete resistors: the resistive behaviour is a side-effect of cell construction and cell parameters.

This drop in voltage as the current increases has four regrettable side effects:

- Heat is generated in the cell which may limit the maximum continuous current
- There is a drop in the efficiency of

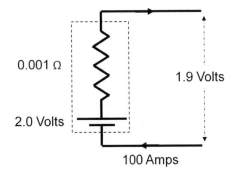

*Figure 6-4 What internal resistance feels like (dotted rectangle is cell boundary)*

the cell (ratio of power out on discharge to power in on charge)

- The available pack voltage drops just when you need it: when power requirements are high

- Cell voltage ceases to be a reliable indicator of state of charge when the battery is being charged or discharged

The last one (cell voltage useless for measuring state of charge when there is current flowing) is a major issue for battery management. Consider for example Figure 6-3. This showed the voltage versus state-of-charge curve for a specific lithium ion cell at a 1 C discharge rate (e.g. 40 Amps for a 40 Amp hour cell, 100 Amps for a 100 Amp

hour cell etc).

Figure 6-5 is Figure 6-3 repeated with the addition of a second voltage versus state-of-charge curve at 0.1 C (4 Amps for a 40 Amp hour cell, 10 Amps for a 100 Amp hour cell)

Note that in this example the terminal voltage at around 95% charge with a 1 C current is almost exactly the same as the terminal voltage at 10% charge with a 0.1 C current. In other words, to a voltmeter an almost fully charged cell of this type at high current is indistinguishable from a largely-discharged cell at low current.

Misunderstandings here have reportedly led

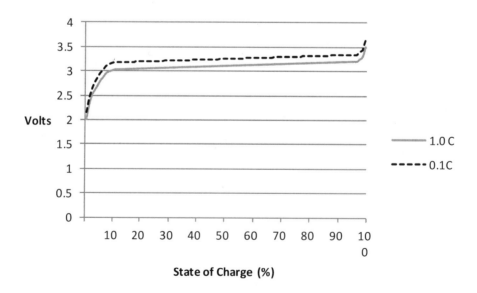

*Figure 6-5 Voltage versus state of charge for a lithium cell at 0.1C & 1.0C discharge rates*

to absurd situations where poorly-designed automated systems have unnecessarily restricted high rate discharges. A low voltage cut off is normally specified for lithium ion cells. Suppose for the sake of illustration we add to the chart of voltage versus state of charge for our typical cell, the curve for a 4 C discharge (400 Amps for a 100 amp-hour pack) - Figure 6-6.

Suppose that for this cell the manufacturer specifies a cut off (minimum) voltage of 2.0, and to add a margin for error, a 2.6 volt limit is going to be enforced. If you look at Figure 6-6, you will see that the 2.6 volt limit will be reached at 50% state-of-charge under a 4 C load. Does this mean that you just cannot use the 4C capability of the cell below 50%

charge?

By no means. The cut off voltage will be specified at a *given discharge rate* (say 0.3 C). At other discharge rates it is meaningless. We will have to come up with another way of figuring out when the battery pack is close to exhausted.

It gets worse. As discussed earlier, different cells in the same pack will have different internal resistances, so even with the same current they will exhibit different voltage drops. A voltmeter is therefore an unreliable indication even as an indication of *differences* in state of charge for different cells in the same pack when the pack is under load.

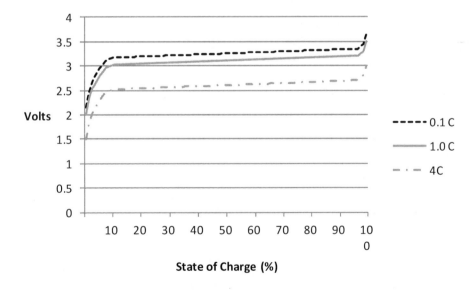

*Figure 6-6 Voltage versus state of charge for a lithium cell with 4C discharge rate added*

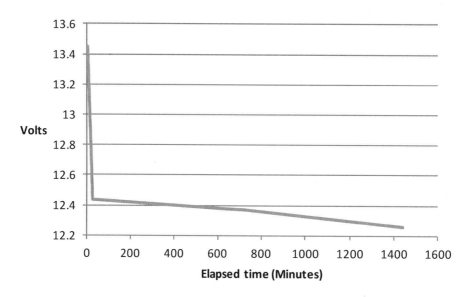

*Figure 6-7 Voltage Settling in Lead Acid battery*

## Complicating factor no 3: Voltage settling

It continues to get worse. It turns out that when a charging current is shut off, the voltage may not stabilise immediately: the voltage will increase short term as the effect of internal resistance goes away, but may then drop back over a period of minutes or hours. Figure 6-7 is an example[8]

The plot starts a few seconds after the charger is disconnected (it was run for just a couple of minutes on a partially-charged battery). Even after 24 hours the voltage had not quite settled back to its fully rested value. The reverse happens on discharge (the voltage bounces back a bit after a discharge current is turned off). This plot is for a lead-acid cell, but lithium ion cells also exhibit voltage changes for a period after current ceases to flow (albeit not necessarily with the same shaped recovery curve).

You could think of charging as being analogous to the delivery of boxes at a warehouse: the charging process is like getting all the boxes offloaded and stacked in the loading bay, but the process isn't wholly complete until the boxes have been moved from the loading bay to the shelves. This is settling.

---

[8] I am indebted to Compass Marine (http://www.pbase.com/mainecruising/battery_s tate_of_charge) for these data

110

## Complicating factor no 4: tolerance to over-charging

Overcharging is leaving the charger hooked up and working after the cell is fully charged. We will discuss this in more detail later in the context of charging single cells, but tolerance to overcharge has a major effect on battery management. Some cell types (flooded lead acid for example) are fairly tolerant to the continued presence of a charging current after the cell has been fully charged. Other cell types (notably lithium ion) are destroyed by single incident of over-charging and may in the process become a fire hazard. Yet others (e.g. valve regulated lead acid – VRLA i.e. Gel and AGM cells) are impaired but not immediately destroyed by over-charging.

Tolerance to overcharge is a helpful characteristic because you can charge a long string of cells together until all are fully charged, safe in the knowledge that the cells that reach full charge earlier than the others will tolerate the ongoing charge current.

## Complicating factor no 5: Tolerance to under-charging

By "under-charging" is meant continuous partial charging, so that the cell never reaches maximum charge. Many (but, anecdotally, not all) lithium ion cell types are tolerant of under-charging. Lead acid cells of all types degrade if treated in this ways due to sulphation (a build up of lead sulphate on the cell's lead plates).

## Complicating factor no 6: Tolerance to over-discharging

That is, running the cell too far down. In extreme cases a weak cell in a large pack may be driven into "cell reversal" (in effect charged the wrong way) as stronger cells continue to drive a discharge current through it when it is exhausted. No cell types enjoy this kind of treatment, and some types are destroyed by it. (The particular problems of large packs are discussed in detail later).

## Complicating factor no 7: Tolerance to under-discharge

Amongst drivers of conventional IC engined cars there is a range of practice in refilling the fuel tank. Some drivers routinely run the tank as low as they dare. Others fill up at weekends, others when the tank gets down to a quarter. Some put in a fixed quantity (say £20 worth) on pay day. There is probably an actuary somewhere who never fills the tank more than ¼ full to minimise the loss of interest on the value of the fuel in the tank.

Similarly, different users of battery-operated power tools or even mobile phones may have different battery-charging habits. Some will recharge only when the battery is exhausted, others will recharge at every opportunity.

In EVs, these charging habits can affect the health of the battery pack. In the context of batteries, "depth of discharge" (DoD) expresses the amount of electrical energy drawn from a cell as a percentage of the energy available from a fully charged cell: so for example a cell at 75% DoD has already delivered ¾ of the energy available from the fully-charged cell. This is analogous to a fuel tank that has been run down to the ¼ full mark[9].

Lead acid and lithium ion cells appear to thrive on under-discharging. One lithium ion cell manufacturer claims a cycle life of 5000 at 70% maximum DoD on each cycle, but only 3000 at 80% DoD. This is a difference in lifetime energy delivery of around 45%. In other words, it makes sense to buy a bigger battery pack than you need and use a smaller percentage of its capacity on each cycle. The longer life will more than pay back the higher cost. The picture is more complex for nickel-based cells. It used to be received wisdom that some types of cell should be fully discharged before recharging. The so-called "memory effect" is said to have arisen from NASA experience with nickel-cadmium cells in satellites although I have not been able to find the original paper. In any case, subsequent experience suggests that whilst this effect occurs, its significance has been exaggerated. One paper for example noted that "...a regular recharge (+ 10% overcharge) restores the usual shape of the voltage-time curve and the capacity..."[10].

---

[9] Confusingly 100% is sometimes considered to be "energy available when new" or the nominal capacity of the cell type, rather than the current actual capacity of the cell in question, (see for example *Electric Vehicle Technology Explained* by James Larminie and John Lowry) so under this convention an old cell that has suffered a 20% loss of capacity will be exhausted at 80% DoD. Not everyone appears to use the term this way however.

---

[10] "The 'memory effect' of nickel oxide electrodes: electrochemical and mechanical aspects" G. Davolio and E. Soragni from Journal of Applied Electrochemistry Vol 28 no 12 1997

# Managing Single Cells (or Monoblocks)

This section will apply the principles we have just discussed (internal resistance, settling, tolerance to over/under charge/discharge) in the context of the management of individual cells, starting with the old familiar lead-acid cell.

## Lead Acid

Most people reading this book will be familiar with the cheap and cheerful battery chargers used to charge the lead-acid SLI (Starter, Lighting, Ignition) battery in a conventional IC car. This is usually just a constant voltage source taken from the mains, courtesy of a transformer and a rectifier (Figure 6-8). The alternator charges SLI batteries on the move in much the same way; by applying a voltage just a little above the battery voltage (Figure 6-9). This causes a charging current to flow. As the difference between the battery voltage and

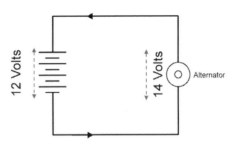

*Figure 6-9 Alternator charging SLI battery*

the applied charger/alternator voltage diminishes, the current falls off to a low value.

You will recall from the chapter on batteries that there are three types of lead-acid battery:

- Flooded,

- Absorbed Glass Mat (AGM)

- Gel.

The second two (AGM and Gel) are often grouped together as "VRLA" (valve regulated lead-acid) batteries. Single lead-acid cells of all types are fairly easy to charge. Regardless of type (flooded, AGM or gel), they share some characteristics which we tend to take for granted because of long familiarity with SLI batteries. Among other things:

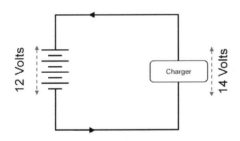

*Figure 6-8 Simple Charger for lead-acid SLI battery*

- Rested open circuit voltage is a fairly good indication of state of charge (this is not true of all cell chemistries – compare for example the Lead acid battery discharge curve depicted in Figure 6-1 and the Nickel metal-hydride curve in Figure 6-2).

- They can accept high currents so are not easily charged at a dangerous rate.

These two effects (increasing voltage with increasing state of charge and a high charge capacity) mean that lead-acid cells are not very prone to thermal runaway (accidental overcharging at a dangerous rate causing an accelerating temperature rise). Also, whilst different manufacturers may specify different charging profiles[11] they are tolerant of non standard profiles.

Lead-acid cells do exhibit several limitations though. Firstly if they are not regularly fully charged they tend to "sulphate" that is lead sulphate crystals form on the lead plates. This irreversibly reduces the capacity of a battery. Secondly they exhibit a fairly substantial rate of self-discharge; that is, they lose charge just sitting on the shelf

---

[11] Charge Profiles are discussed later

with nothing connected to them.

The combination of these two means that lead-acid cells left idle for long periods will degrade. The same thing happens to cells that undergo short partial charges without ever being fully charged.

*Flooded* lead-acid cells exhibit another very useful characteristic: they are not seriously impaired by short periods of mild overcharging (they start gassing which in extreme cases creates risk of explosion, and they lose electrolyte, but they are not immediately and permanently damaged by overcharge like lithium ion batteries). Flooded batteries are designed to vent easily if overcharged, and there is usually provision for topping up with distilled water to compensate for any loss of electrolyte.

Valve-regulated (VRLA, i.e. AGM and Gel) cells on the other hand are a bit more picky. Most of them degrade somewhat as a result of overcharging and the effect is cumulative. Even so, one or two mild overcharging incidents in the life of a cell will not destroy it. As a result, charging a single lead-acid cell is not complex so long as it is fully charged fairly regularly, and so long as VRLA cells are not routinely overcharged.

## Lithium ion cells

As discussed in the chapter on batteries, lithium ion cells deliver far more power for a given size and weight than lead acid and have a far longer cycle life (i.e. you can charge and discharge them many more times before they fail). They are however far more expensive than lead-acid cells and are also more sensitive to mismanagement. This makes accidental damage both more likely and the consequences more severe.

The fundamental problem with lithium ion cells is that they are easily destroyed by overcharging or over-discharging. You can turn a lithium ion cell into a very expensive paperweight in a single over-charge or over-discharge incident. In this section we'll consider the overcharge case in more detail. We'll cover over-discharge in the section on the management of whole packs.

If you overcharge a lithium ion cell, the cell voltage will just keep rising beyond the allowable maximum, irreversibly damaging the cell. Figure 6-10 illustrates the results of one test. Charge continued at the 1C rate after maximum charge was reached. Voltage initially stabilised around 5 volts, and then shot up to a peak of 12 volts after just a few more minutes, at which point the

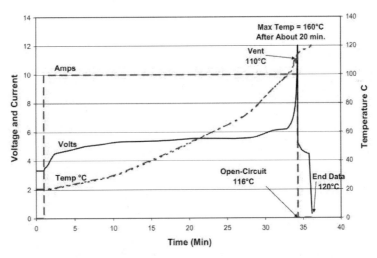

*Figure 6-10 Death-rattle of an overcharged Lithium ion cell (from SAND 2008-5583 Selected Test Results from the LiFeBatt Iron Phosphate Li-ion Battery by Hund TD & Ingersoll D. Reproduced by kind permission of the Sandia National Laboratories, USA)*

cell went to open circuit. It reached a peak temperature of 160 $^{o}$C (hot enough to fry an egg).

This particular cell failed fairly safely, but some lithium ion cells (notably lithium cobalt) are prone to a rather more abrupt termination (i.e. they explode and then burn or vice versa. Try searching YouTube for "lithium explosion").

So much for over-charge. Some lithium ion cells are also fussy about temperature. There are issues at both high and low temperature; the cell manufacturer will probably specify temperature ranges for both charge and discharge. If you allow the batteries to operate outside these ranges you are conducting an experiment rather

than managing your batteries.

High temperatures may impair or destroy a cell with the same finality as over charge or over discharge. Low temperatures are more interesting. Lithium ion cells (like most cell types) lose capacity at low temperature. The paper mentioned earlier [12] records experiments with one lithium ion cell type which indicated around 25% reduction in capacity between 25 $^{o}$C (room temperature) and 0 $^{o}$C (Winter's day in the UK), and that the cell was almost useless at -40 $^{o}$C (Siberian winter). Cells of this type might therefore benefit from some form of cell heating arrangements in a vehicle that was to be used in temperatures typical of a European or North American winter.

Note that this applies to a specific battery type and chemistry, and other types will have different behaviour. One more recent cell type specifies a usable temperature range of -45 $^{o}$C to +85 $^{o}$C.

In summary, lithium ion cells have an operational "window" bounded by temperature and state of charge as

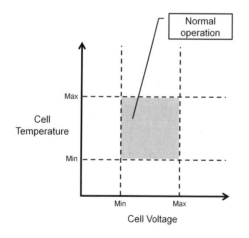

Figure 6-11 Operational window of a lithium cell

[12] SAND 2008-5583 Selected Test Results from the LiFeBatt Iron Phosphate Li-ion Battery by Hund TD and Ingersoll D

indicated by cell voltage at nominal or standard charge/discharge current. If the rested open circuit voltage of an individual cell under these standard conditions gets too high or too low the cell will be damaged. If the cell gets too hot, it will also be damaged. At lower temperatures than recommended it will at least lose efficiency (see Figure 6-11)

The final issue with lithium cells is the maximum current on charge and discharge. These limits are normally specified as C ratings: typically three different ratings, one each for:

- continuous discharge
- short duration pulse discharge
- charge.

So for example a 100 amp-hour cell with a 3C max charge rating implies a 300 Amp maximum charging current.

Managing current limits on discharge is straightforward in principle (if not always in practice): the motor controller needs to know what the limits are and to be programmable to enforce them. The situation on charge is more complex. If you were to use a simple constant voltage SLI charger of the kind described earlier, you have two problems (assuming that the charger voltage is high enough to charge the cell in a reasonable length of time)

Firstly, the initial charge current may exceed the allowable current limit at the start of the charging process (i.e. when the residual cell voltage is at its lowest, so the difference between charge voltage and cell voltage – and hence current – is at its highest). Secondly, the cell will probably be overcharged at the top end.

A modification of this process would be to shut down the charger when cell voltage reached the manufacturers specification. This would solve the top-end-overcharge problem but might (counter-intuitively) not fully charge the cell. Remember that cell voltage as a measure of state of charge is meaningless at the wrong current.

The normal answer to this conundrum is to use a constant-current/constant voltage (CC/CV) *charge profile*. For lithium ion cells this is an initial charge at constant current (voltage rising) in order to charge the cell as quickly as possible within the C rating limit, and then a "settling" period at constant (maximum) voltage with a gradually reducing current, terminated when the charging current falls to a specified level (Figure 6-12). This reduces the impact of

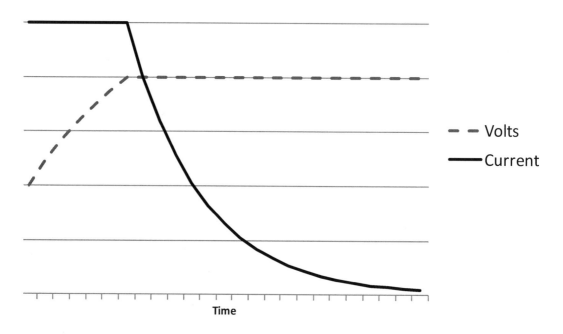

*Figure 6-12 Typical lithium ion charge profile*

internal resistance and charge settling on the charging process.

So the charging system for a lithium ion cell should:

- be programmable to enable it to deliver the correct charge profile, in order to keep both the maximum charging *current* and the maximum cell *voltage* within limits and also to terminate charging when the current has tapered off to a specified value.

- Ideally monitor cell temperatures and reduce charge rate or terminate charging if a cell approaches the maximum allowable figure.

- Ideally provide a backup charge termination mechanism based on elapsed charging time

The second two (temperature monitoring and time-based charge termination) should not be necessary most of the time provided the correct charge profile is used: but cells do sometimes fail and attempting to charge a damaged cell could under some circumstances result in a conflagration.

## Lithium Sulphur cells

There is comparatively little in the public domain about the management of these

118

cells, although there is published work which indicates that they handle overcharge far better than lithium ion. One paper (Safety Performance of Polymer Lithium-Sulfur Cells http://www.oxisenergy.com//downloads/Safety%20Performance%20Polymer%20Li-S_IMLB_2010.pdf) reported that 24 hours of overcharging at the 1C rate caused a trivial temperature rise of under $3^{\circ}$ C. The following do not appear to be in the public domain:

- The effect on cell life of overcharge or over discharge
- The tolerance of the cells to routine undercharging
- The typical state-of-charge versus voltage profile.
- Real world self-discharge rates (there is some evidence that it may be higher than lithium ion)

# Managing entire Battery Packs

Charging a single cell (or a 12 volt monoblock) is one thing, but charging a long string of cells is quite another. Most EV battery packs are charged as a unit by a single charger (Figure 6-13):

As described earlier, the fundamental problem is this: different cells in the string all get the same charging current for the same length of time (they are in series so it cannot be otherwise), but individual cells have different capacities and start with different residual charge (Figure 6-14), so will almost certainly need slightly more or less current/charge duration. The problem gets worse as the string gets longer.

The outcome of this variation between cells is that if the pack is charged to its nominal limit, the weaker ones (those that hold less charge, so need either lower charging current and/or shorter charge duration) will be chronically overcharged. The stronger cells will be chronically undercharged. The impact of this (and potential solutions) vary with cell chemistry, but with many cell chemistries the practical real-world outcome is that:

- The Amp-Hour capacity of a pack

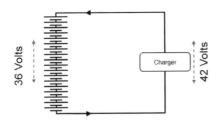

*Figure 6-13 Charging a Long String of Cells*

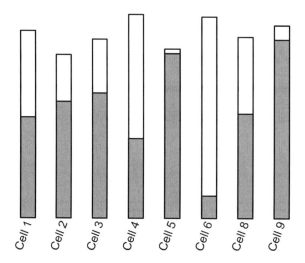

*Figure 6-14 Illustration of cells in a multi-cell pack. Overall bar height represents cell capacity. Shaded area represents current state of charge for each cell*

is, at best, limited to that of its weakest cell

- If the spread of starting states of charge in the pack are too broad, the effective pack capacity will be reduced further

In order to visualise why this is so, imagine a game where each contestant has a tray full of water jugs of varying sizes with varying starting amounts of water. Each jug corresponds to a single cell in a battery pack.

The object of the game is to fill the jugs ("charge" them), carry them across the room and empty them into an urn ("discharge" them). The rules are:

- You can add or take away as much water as you like from the jugs but you must treat them all the same: if you add 100 millilitres (ml) to one jug you must add 100 ml to them all, and if you empty 100 ml from one jug into the urn you must take 100 ml from them all. (This corresponds to charge and discharge currents. Because cells in a battery are in series, the same

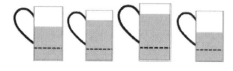

*Figure 6-15 The Water Jug Game*

120

current necessarily flows through each of them on both charge and discharge).

- If the level of the water in any one jug gets below a dashed line marked on that jug an umpire will smash the jug. This is equivalent to going below the minimum allowable State of Charge (SOC) for a cell.

- If any of the jugs overflow (remember you are bound to add the same amount to each), that jug is smashed too (that is breaching the upper SOC limit).

If you were to play this game you would quickly discover two things: Firstly, you have to stop filling all the jugs when the very first jug is full, and secondly, you have to stop emptying them all when the very first jug gets down to the dashed line. So, even if everything else goes right, you are limited by the capacity of the smallest jug. Suppose for example you have nine 1000 ml jugs and just one 900 ml jug. You cannot carry any more water than another player who has ten 900 ml jugs because you have to stop filling all of them when the 900 ml jug fills up, and you have to stop emptying all of them when the 900 ml jug gets down to the dashed line

If one of your jugs starts out too full it will limit you even further: so if (in the preceding example) one of the 1000 ml jugs tops out first, you cannot even fill the 900 to the top.

# Pack balancing/equalisation

The principle of balancing is to add or remove charge from individual cells so that they end up in a similar state of charge. There are in theory several ways to achieve this

- It can be a manual process. At its crudest, an operator can go round the pack at intervals with a voltmeter and a power supply, and when necessary adjust the state of charge to put the cells in line with each other

- Automated "Dissipative" balancing uses cell-level modules to burn off charge from the stronger cells automatically. This necessarily involves dissipating the unwanted energy as heat: hence the term "dissipative". Comparatively low currents work fine so long as the

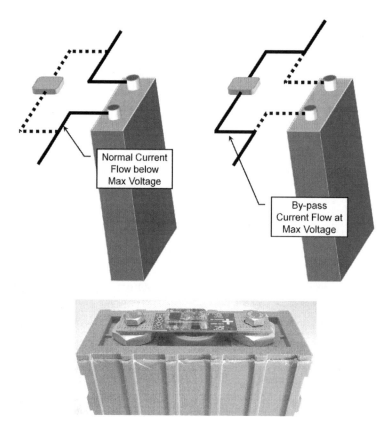

Normal Current Flow below Max Voltage

By-pass Current Flow at Max Voltage

*Figure 6-16 "Shunt" Balancing: top left - normal charging, top right – shunt in operation, bottom – real world Shunt Balancer. Photo courtesy of EV Power, Australia (http://www.ev-power.com.au)*

process runs over a long enough time period.

- A variant of dissipative balancing (using so called "shunt" balancers) divert charge current around the cell when it reaches maximum voltage during charge.

- "Non-dissipative" balancing uses electronics to shuffle power between cells so that they end up balanced.

- Redistribution.

## Manual balancing

The principle here is to manually add or take away charge from high or low cells so that they are all at the same chosen state of charge. The rested open circuit voltage is used as the measure of state of charge. Note "rested, open circuit": it is not enough just to go round with a voltmeter whilst the battery is charging.

## Dissipative & Shunt balancers

The principle of dissipative balancing is to charge the entire string to suit the needs of the cell with the greatest capacity and then to provide a way to dissipate power from, or divert power around, the other cells. Figure 6-16 illustrates shunt balancing. Until the preset maximum voltage is reached, the by-pass circuit is open, so the whole charging current passes through the cell. When the maximum is reached, the by-pass circuit allows just enough current to flow through the cell to keep cell voltage at a preset maximum.

Typically this approach is implemented using small solid-state devices installed across the terminals of each individual cell (or groups of cells in parallel). Most of the time, the balancer is open circuit and inactive. If the cell voltage gets too high during charge, it starts diverting some of the charging current around the cell. These devices are sometimes self contained (no connections other than to the cell itself) and are relatively small, light and cheap. It is perfectly practicable to equip every cell in a high voltage pack with one. However there are concerns over failure modes which are discussed in more detail in the section on automated battery management systems.

## "Non-dissipative" balancing

Non-dissipative balancing essentially robs Peter to pay Paul: a small amount of charge is removed from the cells that are at a higher state of charge and used to top up the other cells. This approach can only really be implemented using relatively complex electronics. However it is done though, the effect is the same. This approach is more expensive: and whilst in theory it is more efficient, losses in the control and "shuttling" circuits can eat into this. Furthermore, the same failure-mode issues arise here too.

## Redistribution

Redistribution in theory allows the maximum power to be extracted from a pack. Essentially each cell is given its own DC-DC converter. This would allow you to continue to draw power out of each individual cell until it was exhausted, using the DC-DC converter to step up from each individual cell to pack voltage (Figure 6-17).

We will not be discussing redistribution further. In an EV it would demand large numbers of DC-DC converters capable of handling large currents, which would become impossibly cumbersome and expensive.

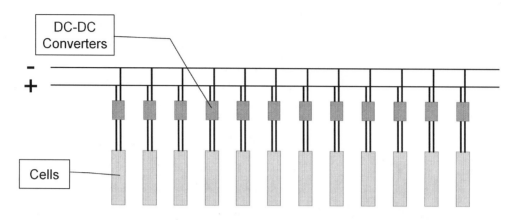

*Figure 6-17 (Theoretical) Redistribution using DC-DC Converters*

## Top Versus Bottom Balancing

So far we have discussed *what* balancing cells in a pack means and *how* to do it. We next turn our attention to *where* to balance them. If you have a pack made up of a number of cells with varying capacities they can only be matched at one point in a charge/discharge cycle. In the real world the capacity mismatch is typically just a few percent, but to illustrate the principle suppose that you had a mixture of 100 Amp hour and 200 Amp hour cells in the same pack, and let's suppose that they were all exactly balanced at (say) 30% state of charge. Now take 2 Amp hours out of the pack. It is no longer balanced: the 100 Amp hour cells are at 28% (2 Amp hours is 2% of a 100 Amp hour capacity) but the 200 Amp hour cells are at 29 % (2 Amp hours is 1% of a 200 Amp hour capacity).

The most usual alternatives are top balancing (ending the *charge* cycle with all cells fully charged) or bottom balancing (all cells run out of power together at the bottom of the *discharge* cycle).

To illustrate this, look at Figure 6-18. Here, the individual capacity of each cell in a 9 cell pack is represented by a bar – the longer the bar, the greater the capacity. The amount of charge remaining in the cell is represented by the shaded area.

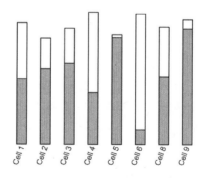

*Figure 6-18 Cell Imbalance*

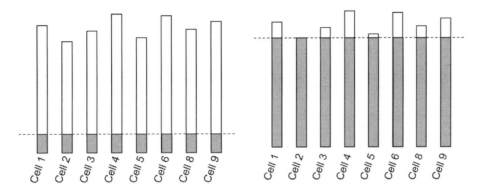

*Figure 6-19 Bottom balanced: discharged state (L) and charged state (R)*

This pack is badly out of balance. For example, cell 5 is almost fully charged and cell 6 is almost fully discharged. A short charge with no balancing on the pack as a whole will quickly top out cell 5 and leave cell 6 charged to only about 15% of its available capacity.

Bottom balancing charges the cells so that at the end of the discharge process they all reach minimum State of Charge (SOC) together (Figure 6-19). Top balancing (Figure 6-20) is the reverse: all cells are charged to maximum capacity. Discharge is stopped when the first cell reaches the maximum allowable depth of discharge.

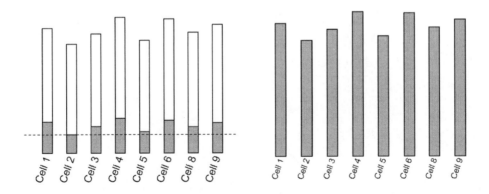

*Figure 6-20 Top Balanced: discharged state (L) and charged state (R)*

It is easier to design an automated Battery Management System (BMS - see next section) to perform top balancing but this is arguably less desirable. Three reasons could be advanced for this:

Firstly the total voltage of a bottom-balanced pack will degrade more sharply near the bottom of the discharge cycle as all the cells approach minimum state of charge. This provides a strong cue that it is time to stop and recharge.

Secondly, top balancing requires that the BMS manage both charge and discharge: it must manage top balancing during charge, and must also tell the motor controller to shut down when the weakest cell hits bottom on discharge. A bottom-balanced pack on the other hand might use a BMS to handle charge and balancing, but the motor controller can be left to handle the bottom of discharge on its own because it can (with a bottom-balanced pack) safely use low overall pack voltage as an indication of depth of discharge (DoD).

Thirdly, as discussed earlier, a cell that is routinely discharged to (say) 70% DoD will last longer than one that is routinely discharged to 80%. Top balancing routinely discharges your weakest cell to the greatest DoD, so by top balancing you are hitting your weakest cell harder than all the others. This will unbalance the pack in terms of cycle life.

Before we leave the topic of top versus bottom balancing, note again that whilst this is a live issue for lithium ion batteries, there is no choice for lead-acid batteries. If lead-acids are balanced at all they simply must be top balanced. Bottom balancing means that the strongest cells are never fully charged: this doesn't matter too much for most lithium ion cells (but do check with the manufacturer). A chronically undercharged lead-acid cell on the other hand will deteriorate due to sulphation.

The behaviour of lithium sulphur batteries in this area is not known. If they are tolerant of overcharging, there is no issue anyway. If overcharging damages them, but undercharging does not, then bottom balancing is an option for them too.

# Battery Management Systems (BMSs)

In this section we will look at automated battery management systems: first of all what they are, and secondly whether they are necessary (or even desirable). Note that this is an area where there is much heated debate, represented on one side by Jack Rickard of EVTV[13] who hates current-generation BMSs for lithium ion packs with a passion, and BMS manufacturers on the other represented perhaps by Davide Andrea of Elithion[14].

## What is a BMS?

In his book on BMSs for lithium ion packs[15], David Andrea suggests six different categories of BMS function. Simplifying this somewhat, there are broadly three types of function that could be performed by BMSs; different makes and models will provide one, two or all three of these functions:

- Monitoring individual cells (state of charge or voltage and maybe temperature) and informing the driver of any that stray outside a defined operating envelope.
- Issuing instructions to other components (such as charger or motor controller) to protect cells based on monitoring data.
- Balancing.

This necessarily involves individual wiring for each cell and/or an individual module for each cell. The exact layout varies. Some BMS's are fully centralised (Figure 6-21). Some use a master/slave topology (Figure 6-22):

BMS sensor wiring does not carry much power, but some wiring complexity is inevitable for either of the topologies illustrated in Figure 6-21 and Figure 6-22. Figure 6-23 shows some of the slave units and wiring of a monitoring system (this installation does no balancing and no control, merely advising the driver of cell voltage and temperature).

---

[13] http://web.me.com/mjrickard/

[14] http://liionbms.com/php/index.php

[15] "Battery Management Systems for Large Lithium Battery Packs" Davide Andrea October 2010. ISBN 978-1608071043

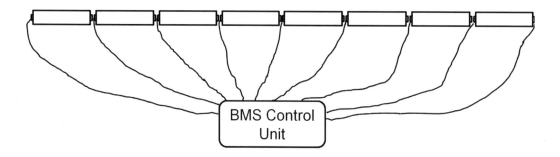

*Figure 6-21 Centralised BMS*

The shunt balancers described earlier are another form of BMS. Figure 6-24 is an example.

Shunt balancers can function standalone, but are sometimes connected to a central unit.

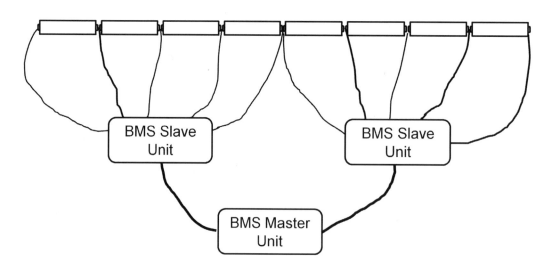

*Figure 6-22 Master Slave BMS topology*

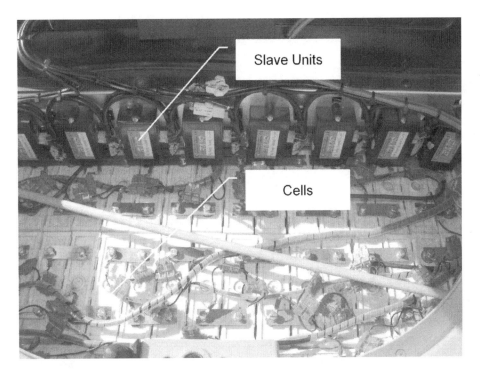

Figure 6-23 Wiring for BMS Slave units

Figure 6-24 Shunt balancers in use. This is the same battery pack shown under construction in Figure 5-1. Photo courtesy of Greg Sievert.

## BMSs for lead-acid packs?

Flooded lead acid battery packs do not need a BMS. Flooded cells can (and should) be equalised by overcharging (undercharged lead acid cells of all types will sulphate). VRLA packs may benefit from some form of automated balancing. There is some evidence to suggest that VRLA packs get out of balance quite rapidly and if this is not regularly corrected the pack dies young. One study of a 48 volt string of AGM monoblocks cycled without balancing found that battery pack failure occurred at about a third of the design life of the individual cells, whilst a similar pack equipped with electronic balancers reached the design life of a single cell[16].

## BMS's for lithium ion battery packs?

My own view is that the problem with BMS's is not what they do when they are working correctly, but what they can do if they go wrong. A BMS for a lithium ion pack is one

---

[16] *Life cycle testing of series battery strings with individual battery equalizers.* Kutkut N.H. Note however that although this paper has been quite widely cited, it was written by a manufacturer of equalisation devices which raises some uncertainty about its objectivity. Note also that this paper applies to AGM cells only.

of a small group of devices which has few safe failure modes. Consider for example the simple shunt balancer:

- If one of the units shorts out, it will melt or even start a fire, possibly damaging the cell or worse.
- If one of the units goes open circuit, that cell will almost certainly be overcharged on the next cycle; this can (as we have seen) result in fire or explosion.
- If one of the units fails to shut off fully, then even if residual resistance is high enough to prevent overheating, the malfunctioning balancer will still bleed charge from the cell and the cell will very probably be over-discharged on the next cycle. This will, at least, kill the cell.

Any of these scenarios is highly undesirable but there is also a statistical issue. If you have 100 cells and if (for the sake of illustration) the mean time between failure (MTBF) of the shunt balancers is 100 years then (crudely) you will lose a cell a year and maybe a car or a house every 5 – 10 years from the resulting fire.

The other aspect of this is the standard of construction. Offhand, the only other bit of electronics I can think of with so few safe failure modes is an autoland system in an aircraft. Autoland systems are built to a high standard with multiple redundancies and are consequently expensive.

Cars are also aggressive environments in terms of vibration, contamination and corrosion (particularly the corrosion of electrical connections). EVs also generate a great deal of electromagnetic interference which can mess with electronic circuits: so as well as having a high specification unit with multiple redundancy, a BMS *and also the BMS wiring* must be designed to operate for years with no failures due to corrosion in the terminals and connectors, no conductor failures due to vibration and no errors from electrical noise. It is a tall order.

Finally a monitoring system may consume some power even when the vehicle is idle, so installing one may cause the very thing that you are striving to avoid. If you forget to charge your EV before you go off on holiday and leave it parked at a low state of charge, the power consumed by the monitoring circuitry may quietly over-discharge the pack.

A BMS capable of handling all of these factors with vanishingly small failure rates would be impossibly expensive. Even then, every wire and every connector in a monitoring or balancing circuit is a potential point of failure, and if you are using many cells, there are many such points of failure outside the BMS itself. Even simple cell-level voltage monitoring in a 100 cell pack is at least 200 connectors and 100 wires. A short circuit in any one of them may mean a fire. An open circuit may be trivial but it might just dupe the charger or controller into trashing the whole pack.

## Alternatives to using a BMS

So is there an alternative to using a BMS? There are broadly two different schools of thought:

1. The first viewpoint is that the cells in a batch are pretty close in capacity and that they can be set to similar states of charge initially; so it is quite acceptable just to use average voltages on charge and discharge, and to add a safety margin to handle the (small) variation between cells. That is, you stop charging well below the safe maximum and stop discharging well before you reach the safe minimum. This is like playing the water jug game knowing that your

jugs are all between 950 and 1050 ml and stopping the filling process when they are all at about the 900 ml mark. The protagonists of this view would argue that in the real world the differences in cell capacities are quite small, and it is cheaper, easier and safer to use slightly bigger cells and charge /discharge more conservatively than to use a complex battery management system that would enable you to get the same end result from marginally smaller cells.

2. The second viewpoint is that it is necessary to monitor the state of charge, or at least the voltage, of each cell individually and continuously. This informs a decision to stop charging when the first cell reaches its upper limit, and stop discharging when the first cell hits the lower limit. This decision can either be made by the user, or by automated control of the charger and motor controller.

On the balance issue there are again two schools of thought:

1. In the real world, lithium ion packs don't go out of balance quickly (if at all) so manual balancing is fine on an occasional as-required basis.

2. The alternative view is that lithium ion battery packs need automated active balancing or the pack will die early

It should be noted that there is little hard experimental data on the extent to which the cells in a lithium ion pack go out of balance as they age. Earlier we cited evidence that AGM packs drift out of balance and die early as a result. I have been able to find no equivalent data for any lithium chemistry. There are several possibilities:

- If lithium (-ion or -sulphur) packs do not in fact drift out of balance then balancing is not required except perhaps some manual balancing when replacing cells or making up a pack from cells in different states of charge.

- If they do drift, but only very slowly, then balancing is something that could be done manually during routine servicing.

- If they drift rapidly then an automated active balancing process is required.

This is perhaps the most significant uncertainty in the management of lithium chemistry battery packs. It will only be settled by research, and we will very possibly find that different lithium cell types behave differently. The research will be difficult given the large number of variables. The long cycle lives of the latest generation of lithium ion cells (2,000 cycles or more) is good news, but it further complicates research. In an EV with a range of 100 miles, 2,000 cycles is 200,000 miles.

In the meantime the arguments will rage based on opinion and anecdotal evidence. Cell manufacturers are probably the best source of data, but speaking personally I would monitor voltages manually at frequent intervals initially. I would however not install a BMS in an EV powered by the current generation of lithium ion cells: BMSs are inelegant, and in engineering, "inelegant" usually means "wrong".

# Estimating remaining charge

For reasons that we have discussed, terminal voltage is not a very reliable indicator of state of charge. Lead acid cells have a relatively linear relationship between rested open circuit voltage and state of charge, but internal resistance and settling time distorts this under load. Nickel based chemistries confusingly exhibit a *drop* in voltage at full charge. Lithium ion cells have such a flat discharge curve that the change in open circuit voltage between, say, 25% discharged and 75% discharged is totally swamped by voltage changes resulting from variations in current.

This is a serious issue: how are we to know how much further we can drive the car before we must recharge it? In other words, what are we going to use for a fuel gauge?

One approach is analogous to a fuel totaliser in an IC-engined car. A totaliser can estimate fuel used and fuel remaining (as well as providing other useful data like instantaneous fuel consumption). It typically zeros itself when the tank is full and then monitors fuel flow rate continuously, keeping a running total of fuel used and deriving fuel remaining by subtraction from the known tank capacity.

The electrical equivalent is sometimes known as a battery monitor and works by counting coulombs rather than litres or gallons. A coulomb is the charge transferred by 1 amp in 1 second (for interest, that's the charge of about 6 billion billion electrons). We have already encountered amp-hours as a measure of

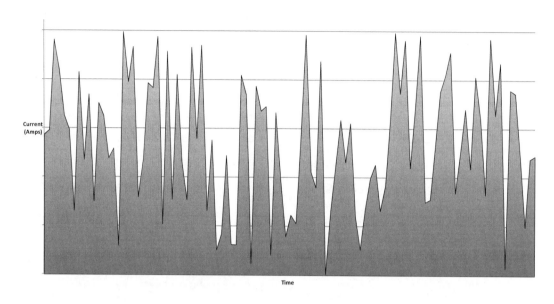

*Figure 6-25 Graph of Current versus Time*

battery capacity (a 1 amp-hour battery will in theory sustain 1 amp for 1 hour), so you could think of an amp-hour as being 60 x 60 = 3,600 Coulombs.

Another way to look at it is that a battery monitor measures the area under the curve of current against time (shaded area in Figure 6-25)

This is not a complex task, requiring an ammeter, a small amount of electronics and a display. In most EV applications, instead of an ammeter in series with the main

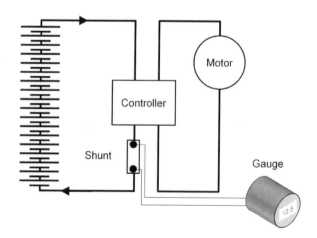

*Figure 6-26 Battery Monitor Schematic*

power circuit, it is easier to fit a shunt in this circuit and measure the tiny voltage drop across it (Figure 6-26). This allows the use of light wiring to the instrument.

These devices are used in several domains – one being yachting and cruising, so several makes and models are available that could be used in an EV. Some of them are even equipped with data ports to support data logging.

There is one issue to consider with any instrument that is connected to the main battery pack: electrical isolation. A normal 12 volt SLI battery has one terminal (these days always the negative) earthed to the chassis. One terminal on each item of 12 volt electrical equipment is earthed so that only a single wire is needed to supply the equipment: the current return path is through the chassis. This is not done with high voltage propulsion packs: the pack is totally isolated from the chassis. This is an important safety issue: if you are working on the battery and touch a terminal accidentally whilst also touching the chassis, no current will flow because there is no return path.

Any sort of instrument directly connected to the pack can compromise this arrangement. If there is a path to ground via the instrument case (for example) then you may be setting yourself up to get a shock from the pack (Figure 6-27)

This problem can be largely avoided by using an isolated current measuring device such as a hall-effect sensor.

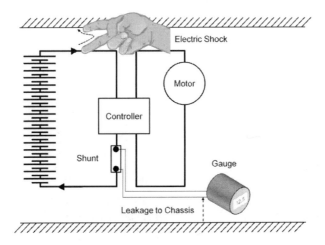

*Figure 6-27 Leakage to the vehicle chassis through instrument creates a shock hazard*

# Fast charging

One of the problems with EVs is the time that they take to charge. This is at least as much an issue of mains supply infrastructure as of the batteries themselves. No charging system can deliver energy to a battery faster than it can pull it from the supply. An ordinary domestic electrical outlet in the UK is capable of delivering about 3 kW: so even if the batteries and charging system were perfect, charging a 25 kW-hour pack from 80% discharged state using such an outlet is going to take between 6 and 7 hours. Even using a 32 Amp industrial plug and socket, charging is going to take at least a couple of hours with a pack of this size. By contrast, there are around 25 kW-hours in 3 litres of petrol which can be pumped into a fuel tank in seconds. Even allowing for the lower efficiency of an IC engine, recharging an EV is clearly a couple of orders of magnitude slower than filling up an IC car.

The reality is that the problem is less significant than might be imagined because most people most of the time don't drive far in a day. So most EV owners recharge at home whilst they are asleep. An EV with an 80 mile useful range would be fine for most of the commuting, shopping trips and school runs that make up the bulk of car usage. A lengthy charging process does mean however that (say) a 300 mile trip in an 80 mile range EV would be impractical in a day with the present charging infrastructure.

The interesting thing about this issue however is that with the right charging infrastructure, EV batteries could be charged rather rapidly. As I write, the maximum charge rate for two common types of Prismatic lithium ion cells is around 3C. By definition, this means that such a cell could in theory be charged in $1/3^{rd}$ of an hour (i.e. 20 minutes). With a 100 Ah cell this means a current of 300 Amps. With a 250 volt pack, and assuming no losses, this would be a 75 kW load. That is about 100 HP and outside the capacity of any domestic power supply. It isn't a silly amount of power however: it is for example well within the range of an uninterruptible power supply (UPS) for a small data centre or the supply for a commercial workshop. If and when EVs become mainstream, equipment and standards will be developed to allow for fast charging on demand.

# Real world chargers

Your choice of charger will be influenced by several factors

- Battery chemistry – charge profiles differ
- Cell capacity (a bigger pack needs a bigger charger)
- Whether you plan to use a BMS and if so, the type of BMS.
- Whether you have high capacity industrial electrical power available

If you don't plan to use a BMS (or if you go for simple passive shunts on the cells) then you may want a charger that can be programmed to allow it to control the whole charging process. You will recall that different cell chemistries require different charge profiles (variation of voltage and current with time). With some chargers, these charge profiles can be manipulated by the user. Others are partially programmed at the factory, shipping with a fixed set of user-selectable charge profiles. Yet others are factory-adjusted for a particular battery pack (chemistry, voltage, amp-hour rating).

On the other hand you may want the BMS to call the shots, and in that case the most important feature of the charger may be that it can be controlled by the BMS. In general if you are going to use a BMS at all, then it is probably best to choose your BMS and then to seek the BMS manufacturer's advice on chargers. If you go this route however, question your BMS supplier about the way in which the charger is controlled, particularly with lithium ion cells. Look for fail-safe i.e. the charger keeps on charging only if the BMS is alive and actively calling for charging to continue, rather than the charger waiting for a shutdown signal from the BMS. If the shutdown signal does not come and the charger keeps charging a lithium ion pack after it should have shut off, then a fire is probable, and will probably occur in your garage in the middle of the night.

## Some typical chargers

Here are some chargers advertised as suitable for both lead-acid and lithium chemistry battery packs when suitably configured. They are mostly about the size of a medium-to-large briefcase and can be installed in the vehicle itself (although in the UK, any onboard charger capacity over 3 kW is wasted as most electrical outlets are limited to 3 kW).

## Brusa NLG511-xxx

3.3 kW ~6.2 kgs

Air or water cooled

Up to 260 Volts DC

Around £2,500 - £3,000

The Brusa is expensive, but highly programmable.

Photo courtesy of BRUSA Elektronik AG (http://www.brusa.biz)

## Manzanita Micros PFC20

4.8 kW at 240 volts input, ~ 7.2 kgs

Air cooled

Up to 360 volts DC

Around £1500 - £2000

Photo courtesy of Rich Rudman of Manzanita Micro

( http://www.manzanitamicro.com)

## Zivan NG3

2.3 kW ~ 6.8 kgs
Air cooled
Up to 312 Volts DC

Around £750
This charger is set up at the factory for a particular charge profile.

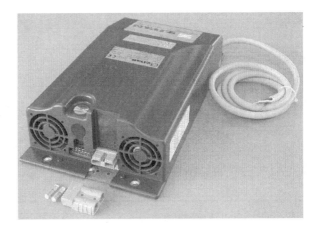

Photo courtesy of KTA Services
(http://www.kta-ev.com/)

## Elcon/Chennic/TCCH

Various models 1.5 - 8 kW, and up to > 400 volts are sold under several different brand names including Elcon, Chennic and TCCH

The 3 kW model weighs around 9.5 kgs. Price range is ~£300 - ~£2,500.

These chargers ship with 10 different charge profiles preset by the factory to suit the battery chemistry and pack voltage.

# In Conclusion

This chapter is probably the most important in the book, and also proved the most difficult to write. With new battery chemistries (such as lithium-sulphur) emerging, much of the detail may change but hopefully some of the principles (e.g. top versus bottom balancing, voltage settling etc) will be useful whatever chemistry you use. Care and thought put into the management of batteries will make a huge difference to the viability of an EV.

One final thought: the Internet is a wonderful source of information but it is also the perfect breeding ground for urban myths and disinformation. There are many armchair theoreticians out there reprocessing ideas of dubious provenance. As ever *Caveat Emptor* – buyer beware!

# 7: Ancillaries

Getting a car rolling under electric power is an achievement, but turning it into safe, legal, reliable and comfortable everyday transport takes more than a traction battery pack, a motor and a controller. Other items that need to be considered include:

- Cabin Heating
- Air conditioning
- Low voltage electric power (lights, horn, wipers etc)
- Power steering
- Power brakes

A major issue here is that the power consumed by these services becomes much more significant as a percentage of total power available. For example, running a 2 kW electrical cabin heater for three hours (a typical running time for a 100 mile range EV) would burn 6 kW hours and reduce the car's range by perhaps 10 – 20 miles. We therefore don't just want to provide these services; we want to provide them as efficiently as possible.

## Cabin Heating

Until the 1960s, car heaters were regarded as optional extras. However since then an efficient arrangement for demisting has become mandatory in most jurisdictions. This requirement is usually met in modern cars by provision of warm air heating onto the windshield. The source of heat is usually the IC engine (although a few cars use electrical windshield heating). More than three quarters of the calorific value of fuel burned in a conventional IC engine ends up as heat rather than useful power. In most modern cars, this heat is normally dissipated by circulating liquid coolant (a mixture of water and anti-freeze) through a coolant radiator. Siphoning some of this heat off to the warm the cabin is trivial – just run some of the coolant through the heater core (a small secondary radiator) inside the car (Figure 7-1).

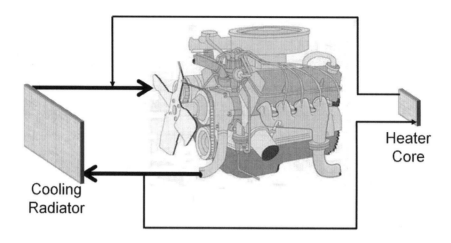

*Figure 7-1 Heater in an IC car*

This approach does not work with electric motors. They are just too efficient. At typical low speed cruise in an EV they don't produce enough waste heat to keep a doll's house warm. There are in practical terms six alternatives:

1. An electrical air heater (akin to a domestic heater) powered from the traction batteries

2. An electrical air heater powered from the 12 volt supply

3. An electrical water heater

4. An air heater powered by gas of the kind used in small boats

5. A petrol or diesel powered air heater

6. A petrol or diesel powered water heater

Each of these has its advantages and disadvantages:

## Electrical Air Heating from the traction batteries

Most EV parts suppliers will sell you a ceramic heater core (Figure 7-2) that you can run from the traction batteries. Power output is typically around 1.5 kW – similar to a domestic electrical fan heater on a low power setting.

This is on the face of it a relatively low cost approach: the heater core itself might be £100 - £200 (although the cost of a suitable high voltage relay should be added to this). Furthermore, heat is available almost instantly. On the other hand, there are several drawbacks.

*Figure 7-2 Ceramic Heater Core. Photo courtesy of EV Source (http://www.evsource.com)*

Firstly, if you are converting a car, and if you want to retain the original heater box, it will be necessary to remove the old heater core and adapt the ceramic replacement to fit in its place.

Secondly, the full traction battery pack voltage must usually be brought inside the cabin. This creates a small but significant extra safety hazard

Thirdly, the heater will take a significant amount of power. Assuming an average heater current draw of 1 kW for an hour in a car that uses 1kW for four miles then at an average of 40 miles per hour (all typical numbers), the heater will reduce range by up to about 10%. To the cost of the heater core itself, it would be fair to add 10% of the traction battery cost.

In a very cold environment it would be possible to use two such cores. It would also be possible to adapt a domestic ceramic heater for this application but this would not be wise unless you know exactly what you are doing.

## 12 Volt electrical Heating

Standalone 12 volt heaters (and 12 volt hairdryers) are readily available (see Figure 7-3 for an example). They may be helpful for demisting and localised heating, but anecdotal evidence and simple arithmetic suggests that they are unlikely to be very satisfactory for heating a large space. The basic problem is that their power ratings are low: ranging from 100 watts to a maximum of about 300 watts. 300 watts is less than most electric drills and kitchen blenders. Compare this with 1500 watts for a ceramic heater core powered by the traction batteries or 2000 watts for the smallest fuel/air heaters discussed later.

The limitation is the massive current required at this low voltage. 300 watts at 12 volts requires 25 amps. An 1800 watt 12 volt heater would require 150 amps. This would flatten the average 12 volt starter battery in a few minutes and require heavy cabling and a very large and expensive relay.

*Figure 7-3 RoadPro C3606A 12 Volt 140 watt heater/demister.*

*Photo courtesy of RoadPro Ltd (http://www.roadpro.co.uk)*

## An electrical water heater

As noted earlier, most IC-engined cars use liquid coolant and use the waste heat to warm the cabin. One approach therefore is to retain the existing heating system, and merely to add an in line electrical water heating element to replace the IC engine water jacket (Figure 7-4, Figure 7-5.)

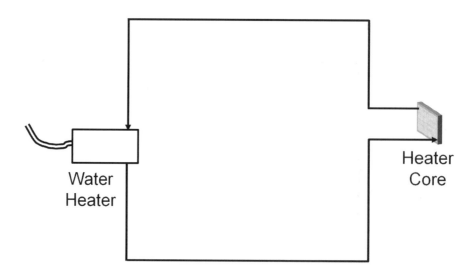

*Figure 7-4 Electric Water Heater Schematic*

*Figure 7-5 In line Water Heater.*
*Photo courtesy of EV Source*
*(http://www.evsource.com)*

As with air heaters, it is in theory possible to draw power from either the traction batteries or the 12 volt system. However the 12 volt option can be discounted on the grounds that at a practical power output the time taken to produce perceptible heat at the heater vents would be very long. Traction battery powered water heaters are a possibility however. They are easier to install than a ceramic air heater and keep high voltages out of the cabin. The high power drain will have the same negative impact on range however. A water heater is also more expensive than an equivalent air heater.

The biggest drawback of any water heater is the time it takes to warm up – probably longer than the heater in a conventional IC-engined car. It has to heat the water in the whole system to a temperature well above ambient before any heat will filter through to the cabin.

## An air heater powered by gas

It is possible to heat your car with the kind of small gas heater that is used in boats and caravans (Figure 7-6). This of course requires a gas bottle which must be

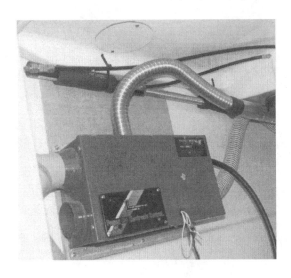

*Figure 7-6 Propex Gas Heater. Photo courtesy of Björn Sjöling and "Restless"*

recharged or replaced periodically. This in turns adds weight and requires space. It also creates a potential safety hazard (although some IC engined cars are themselves driven by natural gas so the hazard isn't novel). Figure 7-6 shows a Propex heater with an output in the range 1.5 – 2 kW. Such a heater requires a 12 volt supply for control and ignition but draws far less power than an electric heater. Note also that it contains a heat exchanger with an external vent. A free standing self contained gas heater in a car would be an accident looking for a place to happen.

These heaters are not cheap either – in the range £600 - £900 new at the time of writing

## A petrol or diesel air heater

Lorry cabs sometimes include overnight accommodation for the driver and require heating. Small motor vessels also need heat when moored. Even some small twin-engined aircraft use heaters that are independent of the engines because of the difficulty of ducting hot air from the nacelles on the wings.

For these applications, several companies have developed heaters powered by petrol or diesel fuel. They obviously use some fuel, but far less than would be required by the main engine.

Once again though, these heaters are not cheap, but they are expressly designed for

*Figure 7-7 Webasto Airtop 2000 ST Air Heater. Photo courtesy of Webasto*
*(http://www.webasto.co.uk/home/en/html/homepage.html)*

vehicle heating. There are, at the time of writing, three main manufacturers of such units: Webasto, Mikuni and Eberspacher. Figure 7-7 is an example of a Webasto heater. Cost for this kind of unit is typically £700 upwards.

Clearly in a pure EV, a petrol or diesel heater has the major drawback of requiring a fuel tank and associated hardware. For some users this goes against the whole principle of EVs. For a hybrid though they look like a good fit. A hybrid will have a fuel system anyway, so tapping into it for a supply for the heater should be straightforward. Fuel consumption is of the order of 1 litre every four hours which is trivial compared with fuel consumption of a normal car engine.

## A petrol or diesel powered water heater

This is an alternative to an air heater. The principle is, once again, analogous to the electrical water heater. The unit is inserted into the existing heater circuit, but instead of being heated by electricity, the heat comes from burning petrol or diesel fuel. As with the electrical water heater, heat is not instant from cold because the water in the system must be warmed up before heat becomes available (Figure 7-8).

## Summary

Apart from the obvious considerations of cost and safety, probably the two biggest issues are power (at least 1500 Watts for a UK winter) and speed of warm up. This tends to rule out both the 12 volt solutions

*Figure 7-8 Eberspacher Hydronic Water Heater. Photo courtesy of J. Eberspaecher GmbH & Co. KG (http://www.eberspacher.com)*

and those relying on water heating (although water heating might be satisfactory with very high powered heating elements).

# Air conditioning

In the average Northern European summer, air conditioning has to be an optional extra. If you must have it, then you are probably going to need professional help, and you must accept that using air conditioning will have an impact on range.

## Conventional Automotive Air Conditioning

Air conditioning rests on three basic principles:

- If you compress a gas it heats up (feel the barrel of a bicycle pump that has been used for a while)

- You can liquefy a gas by compressing it sufficiently

- When a liquid turns into a gas it takes in heat (Sweating illustrates this. When we get hot, we sweat liquid water. This turns into water vapour by evaporation and takes in heat as it does so, cooling us down. An Aerosol can cooling down in use is another example.)

So the air conditioning/refrigeration cycle starts by compressing a suitable gas to make it liquid. This heats the liquid, so it is cooled back down to room temperature. Then the pressure is taken off the liquid, allowing it to expand back into a gas, taking in heat.

A typical air conditioner consists of a loop of tubing with a radiator at each end, a pump between the radiators on one side and an orifice on the other. The whole is filled with a "refrigerant". This is a gas at room temperature and low pressure but becomes a liquid relatively easily when compressed. It works as follows.

The pump (or "compressor" – normally belt driven from the engine in an automotive context) takes in gas at around ambient temperature and compresses it, which heats it up. It is still a gas at this point, but only because it is hot. The hot gas circulates through one of the radiators (the "condenser" – outside the cabin) which cools it back down to near ambient temperature at which point it becomes a liquid.

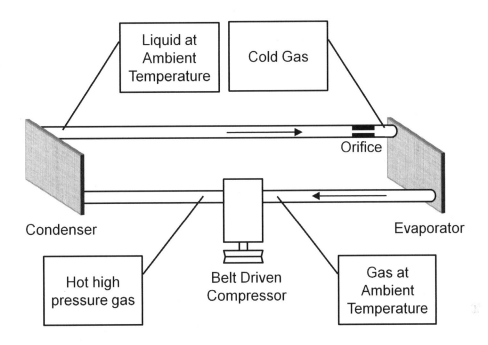

*Figure 7-9 Air Conditioning in a Conventional Car*

The cooled liquid then expands back to a gas as it passes through an orifice, cooling further as it does so. The cold gas is circulated through the other radiator (the "evaporator" – in the cabin). Air is blown through the evaporator by a fan, which causes cool air to circulate in the cabin. The spent gas goes back to the pump inlet and the cycle is repeated (Figure 7-9).

Engine-driven compressors are normally fitted with electrically-actuated clutches. With the clutch disengaged the air conditioning is off; although the compressor pulley is rotating, the compressor itself is not, so nothing happens. The air conditioning is actuated simply by closing a switch which supplies power to the compressor clutch. The clutch engages, the compressor starts to rotate and the air conditioning starts to operate.

The electrical power to the clutch is usually also controlled by a thermostatic switch, so when the cabin temperature gets down to a preset level the clutch disengages so that the cabin is not overcooled.

## Refrigerants

There are currently three types of refrigerant in common use in domestic (household) air conditioners – R22, R407c and R410a. R22 (Freon) is on its way out: R22 should not even be used for maintenance of old systems in the European Union.

At the time of writing, most current automotive air conditioning systems use R134a. R12 was used in the past but has been implicated in ozone layer depletion and has not been used for a while now. Even R134a will not be permitted in brand new designs in the European Union from 2011 as there is concern about its supposed impact on global warming. Some investment has been put into using $CO_2$ (also known as R744 in this context) instead, but $CO_2$ systems run at very high pressures and lower pressure alternatives may be developed: R1234yf is another option.

Automotive air conditioning systems normally use a mixture of lubricating oil and refrigerant rather than pure refrigerant. System rupture (e.g. in an accident) can lead to release of a highly flammable oil mist, although the risk varies widely between refrigerant types. Note also that different types of oil are used in different systems and contamination of a system with the wrong kind of oil can cause rapid failure.

Air conditioning systems operate at high pressures: A typical R134a system might run at 100 – 200 psi. A $CO_2$ system might be 10 times that. The very high pressures of $CO_2$ systems are sufficient to raise safety concerns. Note also that deliberately releasing refrigerant into the atmosphere is illegal in some jurisdictions including the European Union.

All of these issues strongly suggest that messing with air conditioning systems is not a DIY activity. There is nothing wrong with mounting the components yourself but selecting them and charging/discharging should be handled by properly equipped professionals.

## Adapting A/C Systems to EVs

There are at least three ways to adapt an air conditioning system from a conventional car to an EV:

- Replace the engine driven compressor with a self-contained electrically driven one
- Drive the existing compressor with a separate electric motor
- Drive the existing compressor from the drive motor or transmission shaft

In addition there are in theory other options

including

- An ice box
- Thermoelectric (Peltier) devices
- Doing without air conditioning altogether

These alternatives are discussed in more detail below.

## Electrical compressors

At the time of writing, electrically driven compressors are not widely used in cars, although as pure electric production cars such as the Nissan Leaf enter the market they should become available.

Once again, there is an issue related to the power consumption of the compressor. A typical unit in a conventional car may require around 2 Hp (roughly 1.5 kW) which at 12 volts works out at about 125 amps. As with the heater however, it is possible to use power from the traction batteries rather than running a compressor from the 12 volt system. With a 144 volt pack, current would be just over 10 Amps which is far more manageable.

There are a few manufacturers selling electrically driven compressors. Figure 7-10 shows an electrically driven compressor capable of operating from an EV traction

*Figure 7-10 Masterflux® Sierra Air Conditioning Compressor (Photograph: © 2010 Tecumseh Products Company, all rights reserved. Masterflux® and Sierra™ are trademarks of Tecumseh Products Company. http://www.masterflux.com/)*

151

pack (150 – 300 Volts). Models from this manufacturer are available operating at up to about 1.8 kW (6 Amps at 300 volts). They are driven by a brushless DC motor which needs its own controller: a scaled-down version of the main traction controller. Compressor and controller of this sophistication from any source will not be cheap however.

There is theoretically one lower cost source of high voltage air conditioning compressors – domestic air conditioners designed to operate off the mains. It might (I stress "might") be possible to use one of these driven by a DC - AC inverter. However the engineering required to make something like this work reliably would be considerable.

There are a number of considerations. Firstly domestic goods are built to a lower specification than automotive items because they don't have to put up with constant motion and vibration. If you look at the wiring in your cooker, washing machine or refrigerator, you'll see that it is usually less well supported than the wiring under the bonnet of a car. At the least such unsupported wiring in a moving vehicle is likely to cause irritating rattles. At worst the constant chafing and flexing will result in

short circuits, open circuits or both.

Secondly there are potential compatibility issues. This includes things like hose fittings and connectors, refrigerant types and lubricant. As described above a domestic air conditioning compressor might be designed to work with, say R410 rather than R134a and/or require a different lubricant type.

## Separate electric motor drive

This is a practicable, although slightly less elegant solution: obtain a suitable small (1 – 2 HP) electric motor and use it to drive the existing air conditioning compressor. See Figure 7-11 for a well-used example.

This approach typically takes up a bit more space than a self-contained electric compressor and will usually involve more metal-bashing (e.g. the construction of suitable mounting brackets). But it has the advantage of leaving most of the air-conditioning system undisturbed, and should work out to be cheaper than a new compressor. The motor will however need selecting carefully, in terms of power, direction of rotation, supply voltage and r.p.m. A direct drive of the kind shown in Figure 7-11 requires a suitable coupling, although belt drive could be used instead. In addition the motor will need its own controller.

*Figure 7-11 Compressor Powered by External Motor. Photo courtesy of Victor Tikhonov*

## Transmission driven

Perhaps the cheapest solution is to drive the existing compressor from the traction motor or from some part of the main transmission. If your main propulsion motor has an accessible shaft at both ends then it should be fairly straightforward to belt drive the compressor from the unused shaft. The installation issues are similar in scope to those encountered when using the separate electric motor drive described above: there may be significant effort involved with sourcing or fabricating suitable brackets and couplings or pulleys.

The biggest disadvantage of this approach is the obvious one that the air conditioning will only operate when the car is rolling: and even then may not be very effective at low speed. This is a pity as the worst case scenario for most air conditioning systems is slow moving nose-to-tail traffic.

## An ice box

This is a cheap-and-cheerful solution, but one that requires additional preparation for each journey: take along a box of ice in an insulated chest and blow air through it to cool the cabin. It sounds over-simple, but anecdotal evidence suggests that it works. Domestic ice – boxes with fans are available on line for £50 or so (try e-bay or

Amazon). They are going to be bulky but might do the job, especially in a temperate climate where they are only needed occasionally.

A possible approach which is purely theoretical (because I have not seen it implemented) would be to use a portable freezer unit (many models are available). The freezer would be filled with suitable water containers (i.e. ones that could be frozen without splitting). When the traction battery was charging, the cooling unit in the freezer would also be connected and running. Once on the move, the "stored cold" in the frozen liquid could be used to cool the cabin.

## Thermoelectric devices

Thermoelectric cooling normally makes use of the Peltier effect: passing a current across the junction between two dissimilar metals causes one to warm up and the other to cool down - Figure 7-12. [This is the exact opposite of the Seebeck effect exploited in thermocouples]. Peltier devices are used for spot cooling in some situations but are little use for cooling large spaces.

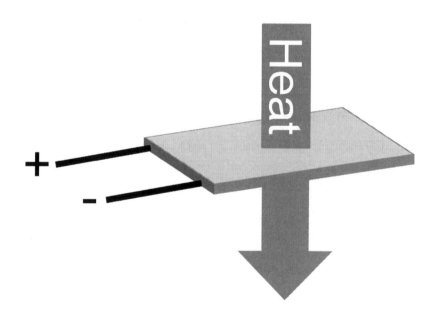

*Figure 7-12 A Peltier-effect device*

## Doing without air conditioning

In the UK, until comparatively recently, only executive barges were equipped with air conditioning. Real men (and women) driving Real cars wore shorts and tee shirts in hot weather, rolled the windows down and, if necessary, sweated. The author drove for years in Africa in air temperatures up to 45 $^{o}$C without air conditioning.

Open windows do increase aerodynamic drag but probably not by very much at typical EV cruising speeds. So one air conditioning option is no air conditioning at all. Strategies for keeping cool without it include the obvious one of not leaving your car to hot-soak in the sun. Park under a tree if there is one handy.

The other option worth investigating is window tint or film. There are strict laws in the UK about the degree of tinting permitted forward of the B pillar (i.e. the back edge of the front doors) but there are films/coatings which claim to reduce solar heating without excessive visual darkening.

So much for heating and air-conditioning. Next we will move on to the 12/14 volt supply normally provided for us in an IC engined car by a belt driven alternator.

# Low voltage electrics (lights, wipers etc)

Virtually all IC-engined cars have a low voltage system (typically a 12 volt SLI battery and a 14 volt belt-driven alternator) for running all the ancillaries needed to make a car roadworthy and to meet legal requirements. Virtually all EVs retain this 12 volt system. It would in theory be possible to swap all the 12 volt ancillaries over to run at traction pack voltage. It would be complex and expensive though: 144 volt headlamp bulbs or 300 volt rear screen heaters are not generally held in stock by your local motor factors.

There is a further complication: a "12 volt" system in an IC engined car doesn't actually operate at 12 volt when the engine is running. The alternator in such a car is normally regulated at just over 14 volts. You can see the difference between 12 and 14 volts if you leave the lights on when you shut the engine down at night; there is a noticeable drop in brightness.

As with heating and air conditioning there are several options for implementing a low voltage power supply in an EV. The two main ones are either to fit an oversize auxiliary battery and charge it up whenever you charge the traction battery, or use a DC-to-DC converter.

## Fitting a larger auxiliary battery

You might recall from the chapter on batteries that ordinary SLI (Starter, Lighting, Ignition) batteries fitted to almost all conventional cars are not designed to be deeply discharged. They are quite capable of delivering short bursts of high current for engine starting, but will be destroyed quite quickly by repeated deep discharge.

If you can cope with dim headlights and sluggish wipers, there is an easy option: there are vast numbers of 12 volt "leisure" batteries on the market designed for use in boats and caravans. A 100 Amp-hour leisure battery should keep the headlights and wipers on for a 40 mile trip without recharging. You might not want to keep a rear screen heater or a 12 volt heater on for too long though with this kind of set up.

A variation of this idea is to use a 14 volt battery to make up for the lack of alternator output on the move. A 14 volt auxiliary

*Figure 7-13 Trojan T875 8 Volt Battery. Photo courtesy of Trojan Battery (http://www.trojanbattery.com)*

battery is a little less mainstream than its 12 volt equivalent. A friend of the author used nickel-iron cells in series with an ordinary 12 volt battery; a better solution might be something like an 8 volt battery and a 6 volt battery in series. The Trojan T875 for example is an 8 volt deep cycle battery with a 170 Amp-hour rating at the 20 hour rate (see Figure 7-13). As might be expected from its capacity (2 – 3 times that of an ordinary SLI battery) it is also fairly heavy at 29 kg. In series with a 6 volt traction battery of the kind often used in golf carts, you would have a solid 14 volt battery pack.

Another alternative would be to build your own pack from individual cells. 7 lead acid cells at 2 Volts each would do it, 4 or 5 lithium ion cells (depending on type) would be close too – and lighter; but would

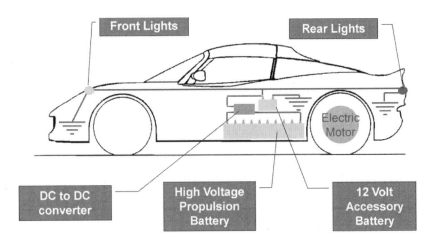

*Figure 7-14 DC to DC converter Schematic*

present some of the same issues of battery management as for a lithium ion traction pack. In either case you may still have a charging issue: chargers for 14 volt batteries aren't very common either.

## DC – DC converter

A DC to DC converter is much like a step-down transformer, except that it works with DC rather than AC. It draws current from the propulsion battery at high voltage and delivers power to the low voltage DC circuit at 12 volts (Figure 7-14). The simplest form of DC – DC converter is a pair of resistors making up a simple dividing circuit (Figure 7-15) not unlike the crude form of variable resistance controller described in an earlier chapter.

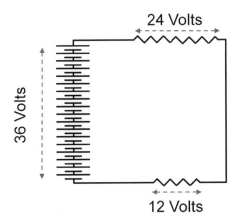

*Figure 7-15 A simple voltage divider*

For much the same reasons as that crude controller, a simple voltage divider is impracticable in an EV. The low efficiency wastes power and generates excessive heat.

There are several different more sophisticated DC-DC converter designs that use some combination of inductors, capacitors and solid state devices. For our purposes buying a DC - DC converter from a reputable EV supplier is probably good enough.

You do have a choice to make however.

Will you use a DC-DC converter on its own as the sole source of low-voltage DC power, or will you retain the 12 volt battery? Retaining the 12 volt battery has the disadvantage of extra weight and potentially adds to your battery maintenance chores. On the other hand, the battery is a buffer against failure of the DC – DC converter and may allow you to select a converter with a lower power output.

Another critical consideration is power when you are parked. Relying wholly on the DC to DC converter for 12 volt power means that if you parked in the street at

| Component | Typical Power Consumption (Watts) | Component | Typical Power Consumption (Watts) |
|---|---|---|---|
| Headlights | 120 | Instrument and interior lights | 10 |
| Front spot/fog lights | 100 | Ventilation fans | 25 |
| Front Wipers | 50 | Audio equipment | 50 |
| Rear Wiper | 25 | Seat adjustment motors | 50 |
| Tail lights | 10 | Front and/or rear screen demisters | 250 |
| Brake lights | 42 | Power windows | 25 |
| Rear fog light | 21 | Central Locking | 10 |
| Number plate lights | 20 | Electric Power Steering Pump | 750 |
| Hazard Warning/indicators | 84 | **Total** | **1642** |

*Table 7-1 Typical Power requirements [1642 watts at 12 volts is > 100 amps]*

*Figure 7-16 MES DC to DC Converter. Photo courtesy of MES s.a. (http://www.mes.ch)*

night you would need to leave the traction pack live. Moreover if the DC – DC converter drains power from the traction pack even when no low voltage power is being used, then you increase the risk of damaging the traction pack if you forget to shut it down and leave the car standing for a long period.

If you decide to dispense with the 12 volt battery and rely on the DC – DC converter alone, you will need to select a converter that can handle peak power requirements – in other words, the converter must be capable of sustaining rated voltage with all electrical equipment switched on – imagine a dark, wet and cold night with the heater fan, screen demist, wipers, side, head, and fog lights, hazard warning etc all on. Table 7-1 lists typical power requirements for components. These can vary very widely,

so you will want to check the components in your car. This suggests that a low power (150 – 250 Watt) converter isn't going to keep the car supplied with power on that wet November night.

If you decide to retain a 12 volt battery then you only have to handle average rather than peak load which should be a lot less. If you have your screen heater(s) on a time switch for example, you may be able to get away with a converter operating at a few hundred watts.

A 1 - 2kW DC to DC converter like the one shown in Figure 7-16 might cost £1000 as opposed to £300 - £400 for a 400Watt unit. Weights for these larger converters are in the range 2.5 – 5 kgs. They are neither bulky nor heavy.

# Power steering

There are three forms of power-assisted steering in common use in IC-engined cars:

- Conventional hydraulic assistance powered by an engine-driven pump (usually via a V belt). Valves attached to the steering column direct hydraulic pressure to one or other side of a ram, which assists the driver in turning the wheels.

- Electro-hydraulic systems – similar to belt driven, but the hydraulic pump is powered by an electric motor rather than the engine.

- Pure electric systems – no hydraulics. Assistance is provided by what amounts to an electric motor controlled by sensors that read the torque being applied to the steering column.

If your car already has a pure electric or an electro-hydraulic system you will probably just want to retain it unmodified. Such systems do consume substantial power however, and the system will almost certainly be 12 volt: which means that you will have to size your DC – DC converter and/or 12 volt battery to handle the substantial extra load.

Note that some power steering systems incorporate a small cooling radiator: the power consumed by the pump has to go somewhere, and much of it ends up heating the power steering fluid.

If the system in your car uses a conventional engine-driven hydraulic pump then you have several alternatives (much like those associated with air-conditioning):

1. Replace the engine driven pump with a 12 volt pump
2. Use an auxiliary 12 volt motor to drive the existing pump
3. Replace the engine driven pump with a pump powered by pack voltage
4. Use an auxiliary motor operating directly from pack voltage to drive the existing pump

These are described in more detail below:

## Replace the original pump

The most likely tactic here is to use an electrical power steering pump from a different car. This will, almost by definition, require a 12 volt supply. The pump from the Toyota MR2 has been used in some conversions (see Figure 7-17)

*Figure 7-17 MR2 Power Steering Pump. Photo courtesy of EVDrive*

*(http://www.evdrive.com/)*

One thing to watch out for here is the control system used for the pump. If the pump just runs continuously you are using electrical power from your DC-DC converter and/or your auxiliary battery to no benefit for the 99% case where you are just rolling down the highway needing negligible assistance.

## Re-use the existing pump

That is, drive the existing belt driven power steering pump with a small DC motor either directly or via a belt drive. This is more or less exactly the same principle as using a separate motor to drive the air conditioning compressor (see Figure 7-11)

As above though, you have a choice to make: will you leave it running all the time that your main contactor is closed, or will you only run it when it is actually needed? The former is simpler, but the latter more efficient: a 1.5 – 2 HP electric motor running continuously is going to have a significant impact on range, and will affect every journey, not just the hot or cold weather trips as would be the case with heaters and air conditioners.

*Figure 7-18 MES Vacuum Pump Photo courtesy of MES s.a. (http://www.mes.ch)*

# Power brakes

This one is relatively simple. Most modern cars (not all) have power assisted brakes. This usually requires a source of low pressure air (often loosely described as "vacuum" although it is nowhere near a vacuum in reality).

Almost universally, petrol engined cars use a tapping in the engine induction system as their low pressure source. However some high performance petrol engined cars and some diesel engined vehicles use an electrically driven "vacuum" pump. This means that there is a plentiful supply of pumps available for use in EVs: although once again, you will need to ensure that you take account of the power drain in your sizing of DC - DC converters, wiring etc. Figure 7-18 illustrates a vacuum pump made by MES DEA. Such pumps are relatively light and compact: the MES pumps are around 1.3 – 1.4 kg depending on model. Be aware that some pumps used on diesel vehicles are engine driven rather than electric.

# In Conclusion

Not many years ago, cars were ticklish devices. Until quite recently it was normal to come back from a long road trip with one or two things to fix. One car the author owned in the 1980s consumed three alternators, two water pumps and a range of other electrical and mechanical components over a three year period. In the 60's and 70's it was commonplace for radiators to spring leaks, cooling hoses to split, fuel tanks to rust through, handbrakes to freeze up, master cylinders to fail, starters and starter contactors to jam or fail, carburettor jets to clog up: even window winders failed, making it impossible to shut the window. Seats were often uncomfortable on long trips, heating inadequate and (in Europe at least) air conditioning was the preserve of the wealthy.

Modern production cars are by contrast very usable and very reliable: we get used to jumping into them and driving hundreds of miles without even checking the oil. Breakdowns are rare events. Many (perhaps most) modern cars will never leave their driver stranded by the roadside in their entire working lives. This taken-for-granted reliability is not a natural state of affairs. It is the result of great attention to detail and the judicious exploitation of developments in design, materials and manufacturing techniques. It borders on a modern engineering miracle.

This sets a high bar for conversions: if your EV has 1970's standards of comfort and reliability you will use it less than you would if it were more reliable. If it is snowing outside and you have the choice of a snug, warm and well-insulated IC car or an EV with marginal heating and a tendency to leave you stranded by the side of the road, which are you going to take? Brakes, steering, heating and cooling are important: if you convert a car that was designed and built to use servo assisted brakes and power steering, simply dispensing with them may not be an option on grounds of safety or usability. If your arrangements for heating and/or air conditioning are inadequate for your local climate you may not be able to live with it year round and it will become a fair-weather toy, or at best used for only part of the year. If the 12 volt electrics on your EV are marginal you may be nervous about taking the car out at night or in the rain.

If you want an EV that is usable everyday transport rather than a proof-of concept testbed, you need to put the same level of effort, thought and investment into these areas as you put into the drivetrain.

# 8: Other Technologies

In Chapter 1 we promised that we would look at alternatives to batteries as a replacement for petrol or diesel fuels in vehicle drivetrains. These alternatives include

- Fuel cells
- Flywheels
- Hybrid power trains
- Ultra-capacitors
- Compressed Gas Energy Storage
- Superconducting Magnetic Energy Storage
- Inductive charging on the move
- Overhead power

In this chapter we examine some of these ideas, albeit briefly. Many of them are of purely theoretical interest either because parts are not available or because a suitable infrastructure does not exist, or both.

## Fuel Cells

Fuel cells are an interesting technology, but are, at least in the short term, a monstrous red herring.

The concept of a fuel cell is simple. A liquid or gaseous fuel is used to create electricity directly, and this drives an electric motor. The usual fuel is hydrogen. An onboard hydrogen tank is topped up at a filling station from a special pump in much the same way as an IC engine car fuel tank is filled with petrol. On the move, the hydrogen is piped to a fuel cell which combines the hydrogen with oxygen from the air to produce water, and in the process generates electricity which is used to drive an electric motor (Figure 8-1).

A vehicle powered by a hydrogen fuel cell would be as quiet as one powered by batteries, and its exhaust emissions would be just water; so fuel cell cars are, like EVs, low noise, and pollution-free at the point of use. In addition, there would be none of the

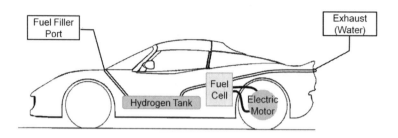

*Figure 8-1 Hydrogen Fuel Cell propulsion*

problems of extended recharging time or excessive weight that you can get with battery-powered vehicles.

Fuel cells however have several problems. The first issue is the source of the hydrogen itself. Fuel for IC-engined cars starts as oil extracted from the ground, piped or shipped to a refinery. The refinery produces the petrol and diesel required which is then shipped again to a network of filling stations where it is bought by consumers (Figure 8-2)

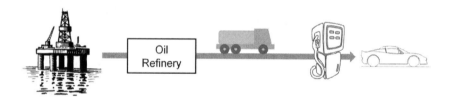

*Figure 8-2 Conventional Fuel supply chain*

Current industrial scale production of hydrogen uses *natural gas* as its raw material; so the supply chain is virtually identical in principle (Figure 8-3). So, whilst fuel cells would give you the town-centre benefits of low noise and low pollution, they

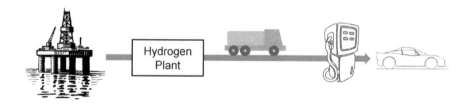

*Figure 8-3 Hydrogen Fuel Cell Supply chain*

would do little to tackle the fossil fuel usage issues other than by a possible improvement in the "well to wheel" efficiency.

It is true that hydrogen could be produced by electrolysis of water which might be achieved using electricity from renewable sources. This does not pass a simple sanity check for mass market application: why generate electricity, use the electricity to produce hydrogen, ship the hydrogen around the country in tankers and finally generate electricity again from the hydrogen, when you can just use the electrical energy directly in a battery electric vehicle?

Secondly, the technology is immature. There are no mass-market applications for fuel cells. By contrast, Lead-acid traction batteries have been used in things like fork lift trucks for many years, and even the newer cell chemistries like nickel metal hydride and lithium ion are widely used in consumer electronics. Fuel cells have turned up in a bunch of prototype and concept vehicles, and some in edge-cases like military submarines and spacecraft; and that is about it.

There have even been doubts expressed about the availability of the platinum used as a catalyst. One 2008 paper estimated that platinum availability might restrict fuel-cell vehicles to about 15% of the market[17]. A TIAX report for the US Department of Energy was more bullish but still concluded that "The platinum industry has the potential to meet a scenario where FCVs [Fuel cell vehicles] achieve 50% market penetration by 2050, while an 80% scenario could exceed the expansion capabilities of the industry..."[18]

Thirdly the use of hydrogen fuel cell vehicles would require a vast new infrastructure for the large scale production, transport, storage and dispensing of hydrogen. Every filling station would need a hydrogen "pump" or refuelling point. We would need new hydrogen production plants and a new tanker fleet. Furthermore, this infrastructure has to be in place for the

---

[17] Platinum Supply and the Growth of Fuel Cell Vehicles by Justin Boudreau, Eugene Choi, Oljora and Rezhdo, Khalid

[18] Platinum availability and economics for PEMFC commercialization," Tiax LLC report D0034 to the US

Department of Energy, Cambridge, Massachusetts, December 2003

very first hydrogen fuel cell vehicle to be viable: if you are going to drive across the country, there have to be hydrogen filling stations across the country.

Battery vehicles on the other hand need no new infrastructure. The core infrastructure (the national grid) already exists. To drive an EV anywhere in the world today, all you need is an electrical extension cord (provided that you have an onboard charger, which most EVs do). EVs are also incremental. Scaling electricity production to handle mass-market EVs would be a matter of economics and planning around a well-understood industrial process.

There have, nevertheless, been a stream of fuel cell "concept" and "demonstrator" cars from the motor industry over the last few years, and the concept is often touted as the long term answer. A cynic might be forgiven for thinking that concept cars based on an immature technology and on a fossil-fuel supply chain might be an exercise in foot dragging: an excuse for failing to make substantive progress on usable production cars based on an imperfect but entirely practicable battery technology.

# Flywheel Energy Storage (FES)

Instead of storing energy in a battery or using a liquid fuel that gives up energy, it is in theory possible to store kinetic energy (the energy of motion) in rotating mass – a flywheel. Flywheel Energy Storage (FES) is an idea that has been around a while: most men over 40 (and quite a few under 40) have used FES devices. That was the principle behind countless toy cars that were fitted with flywheels geared to the driving wheels so that the flywheel rotated very rapidly when the toy was pushed along (see Figure 8-4)

The same principle is used in some F1 cars: a flywheel is accelerated during

*Figure 8-4 Zecar Flywheel-powered toy.*
*Photo courtesy of Justin Ketterer*
*(http://justinketterer.com)*

braking, absorbing energy which is released to boost subsequent acceleration.

This technology is not a joke. A ring weighing 1 kg and 0.3 metres in diameter rotating at 50,000 r.p.m. would (theoretically) store about 300 Kilojoules or about 83 watt hours. The only snag is that the centrifugal force would be about 400 tones. 50,000 r.p.m. is not however absurd. Modern turbochargers run over 100,000 r.p.m. By contrast a 40 Amp hour lithium ion cell would store about 132 watt hours and weigh around 1.5 kg. The development of practical flywheel energy storage is partly a function of advancing materials science.

High speeds are needed, and therefore high centrifugal loads are created. Structural failure of such a flywheel is potentially catastrophic: the rim of our theoretical 300 mm diameter flywheel would be moving at over 0.75 km/second. This is the same order of magnitude as the muzzle velocity of a rifle bullet.

An uncontained flywheel failure would be potentially lethal to anyone in the vicinity, so part of the penalty of an FES system is the weight and expense of the hardware required for containment in the event of structural failure of the flywheel.

# Hybrids

We mentioned hybrids in chapter one. A hybrid in this context is usually a vehicle with both battery electric and IC elements in its drivetrain. Like many of the concepts we have been discussing the idea is not new. In the section on wheel motors we mentioned the Lohner Porsche EV. Dr Porsche extended this idea by adding a generator to create a Hybrid (Figure 8-5)

Hybrids have had a bad press in the EV community, perhaps because the best known and most successful commercially

have been "plug free" hybrids such as the Toyota Prius which ultimately get all their

*Figure 8-5 Replica of the Lohner Porsche Hybrid. Photo courtesy of Andreas Zarini*

*Figure 8-6 BYD F3DM Plug-in Hybrid. Photo courtesy of Anthony Kendall*

power from petrol pumped into the tank. This bad press is a mistake. Unless and until battery technology gets to the point that range ceases to be an issue, a good hybrid may sometimes be worth considering for some applications

Hybrids may also be a way of easing the transition from IC to EV in a commercial and emotional sense: for some buyers the security-blanket of a fill-up-and-go IC drivetrain may be necessary to encourage them to buy into electric power at all.

So what kinds of hybrid are there? There are four classes which you can think of as a 2 x 2 matrix. Some hybrids ("Plug-in" hybrids) are, at their best, ordinary EVs with an extra IC engine just to get you there on the rare occasions that your journey is greater than the EV-mode range. This sort of hybrid can be operated rather like a

sailing yacht with an outboard motor – i.e. with the IC engine used as little as possible. For example, the Chinese-made BYD F3M (Figure 8-6: not available in the West as I write) has an electric only range of around 60 miles – so could be used as an electric vehicle most of the time by many drivers whose daily commute or school/shopping trips are less than this. Such a vehicle, used most of the time in EV mode, might in practice go hundreds of EV miles for every gallon of petrol used.

Other hybrids (like the Toyota Prius mentioned earlier) merely use the electric motor to smooth out the power demand on the IC engine, thus making it more efficient. These vehicles are sometimes referred to as "Plug-free" hybrids (marketing speak for "you cannot charge the batteries from an external power source"). This sounds useless but curiously it does actually work.

*Figure 8-7 Toyota Prius Plug-Free Hybrid. Photo courtesy of Brett Cravaliat*

For example the second generation Toyota Prius (Figure 8-7) was a 5 seater which could accelerate from 0 to 60 mph in just over 10 seconds. It had a 1500 cc petrol engine putting out 76 BHP. The contemporary Volkswagen Golf Hatchback 1.6 S FSI weighed a similar amount, accelerated to 60 in much the same time but had a 1.6 litre engine putting out 113 BHP. The official fuel consumption figure for the Prius was 65 mpg. The Golf, 41 mpg.

The reason it works is that you can use a smaller IC engine and run it closer to maximum efficiency, using the electric motor to make up the deficiency in power during hard acceleration.

Plug-free hybrids are better than nothing, but the improvement in efficiency over a good diesel is marginal. Many people regard them as green fashion-statements, essentially a conventional car with a transmission augmented by an auxiliary electric motor and batteries.

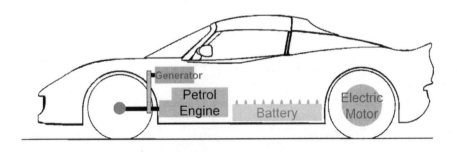

*Figure 8-8 Parallel Hybrid: both the IC engine and the electric motor can drive the wheels*

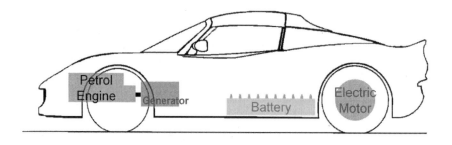

*Figure 8-9 Series hybrid: the IC engine drives a generator and the electric motor(s) drive the wheels*

There is another way to classify hybrids apart from "Plug-in" and "Plug-Free".

"Parallel" hybrids use both the IC engine and the electric motor to drive the wheels (Figure 8-8). By contrast in a "Series" hybrid (Figure 8-9) the IC engine does not drive the wheels directly: instead it drives a generator which recharges the batteries. The batteries provide power to an electric motor which drives the wheels.

Indeed some EV owners have even built separate trailers with onboard generators to tow behind their cars on long journeys (Figure 8-10).

Parallel hybrids are more complex but in theory offer better performance. When you are accelerating hard the control systems can tell the IC engine to forget about charging the battery and help the electric motor lay some rubber. Once you are in the cruise, the IC engine goes back to charging the batteries.

Series hybrids are simpler: conceptually they are an ordinary EV with an onboard

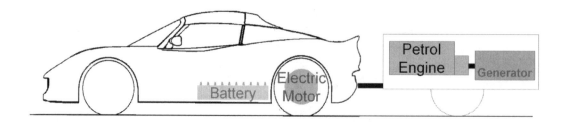

*Figure 8-10 Range-extending trailer*

generator to charge the batteries. A series hybrid doesn't need the complexity, weight and expense of a transmission that can mix power from two power sources. Nor does it need sophisticated control software to orchestrate the output from two power sources with very different power and torques curves. As a minimum all you need to do is crank up an onboard generator when needed. The crazy BBC *Top Gear* trio demonstrated the principle with an absurd EV powered by a couple of lead acid batteries supplemented by an off-the-shelf portable generator.

So there are actually four categories of hybrid "Plug-in Series" "Plug-in Parallel", "Plug-free Series" and Plug-free Parallel" (Figure 8-11).

In the real world, if you want a hybrid, the series hybrid is probably the right way to go. Let the different kinds of motor play to their strengths: IC engines are at their (limited) best running at constant speed: such as when running a generator. Electric motors are just far better at propelling a car.

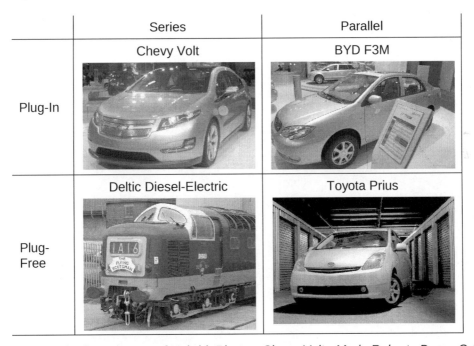

*Figure 8-11 The four classes of Hybrid. Photos: Chevy Volt - Mario Roberto Duran Ortiz; Deltic - Nick Drew; BYD - Anthony Kendall; Toyota Prius - Brett Cravaliat*

One possible parallel hybrid layout that might be made to work is a conversion of a four wheel drive vehicle that retains the IC engine but only driving one set of wheels and adds an electric drivetrain to power the other axle (Figure 8-12).

This, it must be said, is a largely theoretical possibility: it has been discussed and there are a few concept conversions (e.g. MIRAs H4V) that match this pattern, but I am not aware of any daily-use amateur conversions. One of the issues is that you are probably limited to a direct-drive electric motor: two separate gearboxes would be unthinkable.

Another intriguing possibility is the so called 'Through-The-Road-Hybrid' (TTRH) configuration. This involves adding in-wheel motors (like the Protean example discussed earlier) to an existing IC-engined vehicle without modification to the existing drivetrain. This design enables the hybridisation of any existing vehicle resulting in 3 driver-selectable operating modes;

- IC engine only,
- electric only,
- electric + IC hybrid.

Accessories like heating, power steering, air conditioning and power brakes are a concern with any form of hybrid: if the IC

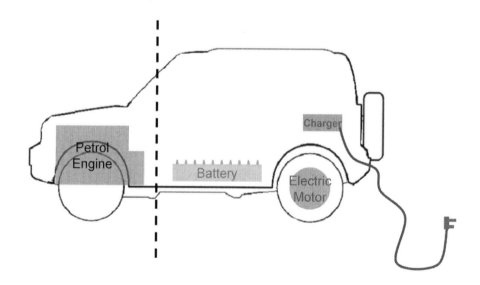

*Figure 8-12 A possible Parallel Hybrid Layout*

engine is not operating much of the time, you will have to modify these components, but then what do you do on that long road trip when the propulsion battery is discharged 50 miles from home? In the end, hybrid drivetrains will always be heavier, more complex, more expensive and higher maintenance than either an EV or an IC-engined car. You are probably better off to sell the IC drivetrain components and use the money to buy a few more batteries which you can install in the space vacated by the engine. Any money left over can be used to hire a car for the couple of occasions a year when you need more range.

# Ultra-capacitors

Small capacitors are commonplace in electronic equipment. Like batteries, they store electrical charge but usually in tiny amounts and in a way that is different to the typical battery. Unlike most batteries, the voltage across a capacitor tends to vary linearly with the amount of charge stored. If we re-use our earlier analogy of an electrical circuit being like the flow of water in a pipe, a capacitor is a bit like a rubber diaphragm across the flow of water: it offers little resistance at first but the resistance builds up rapidly as the flow continues (Figure 8-13). Similarly, as a capacitor is charged up its voltage rises in approximate proportion to the charge stored.

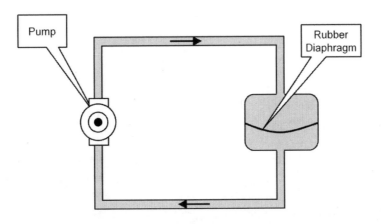

*Figure 8-13 The water-pipe equivalent of a capacitor*

Capacitance (the "capacity of a capacitor") is measured in Farads. The terminal voltage of a 1 farad capacitor would rise by 1 volt for every coulomb (amps x seconds) of charge stored. So a 5 farad capacitor charged to 100 volts would absorb 500 coulombs, or (say) 50 amps for ten seconds. Most capacitors used in electronic equipment are nowhere near as big as this. The range of capacitance in this kind of application would typically be from around .000005 microfarad up to 100 microfarad (a microfarad is a millionth of a farad).

It is however possible to build capacitors with ratings in the thousands of farads. For example a Maxwell BCAP3000 capacitor with a rated capacity of 3000 farad at 2.7 volts weighs around half a kilo and is about the size and shape of a big cylindrical lithium-ion cell (Figure 8-14). It could store about a bit over 3 watt-hours. By contrast a Headway 40160 lithium ion cell of similar volume and weight stores 16 Ah at 3.3 volts or around 50 watt-hours. The Maxwell capacitor is substantially more expensive than the Headway cell.

So an ultracapacitor of this type is not yet ready for prime time as the only source of power for an electric vehicle, but it might conceivably get there. In the meantime they have a niche as a way of boosting the specific power of a battery drivetrain: capacitors can provide a short-term burst of power for acceleration and/or accept a very

*Figure 8-14 Maxwell ultracapacitors. Photo courtesy of Maxwell Technologies, Inc.*

*(http://www.maxwell.com)*

high charge rate for regenerative braking.

Ultracapacitors have a couple of huge advantages over batteries. Firstly the specific power of these cells is staggering. The BCAP3000 can sustain almost 150 Amps continuously and over 2000 Amps for 1 second. That's equivalent to a C rating of 66 continuous and almost 900 peak.

The second great benefit is cycle life. Lead acid cells manage a few hundred cycles; the best lithium ion cells a few thousand. Maxwell claim over a million cycles (albeit at a 50% duty cycle) for their ultracapacitors.

Capacitors are also more forgiving when it comes to management. They don't mind being undercharged, they don't have cell memory and they don't need watering like flooded lead-acid cells. They don't explode when overcharged or give off fumes. Long series strings of capacitors do need balancing however. The Maxwell 125 Volt BMOD0063 modules (Figure 8-15) for example have built in balancing electronics.

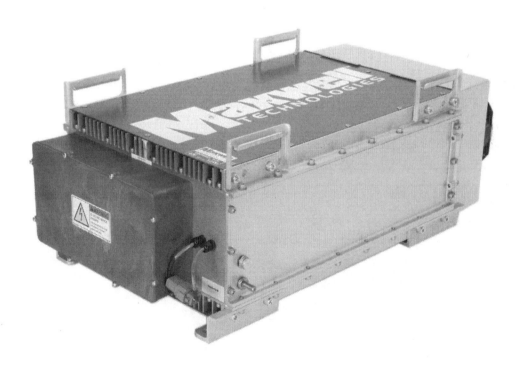

*Figure 8-15 Maxwell BMOD0063 Ultracapacitor Module. Photo courtesy of Maxwell Technologies, Inc. (http://www.maxwell.com)*

# Compressed Air Energy Storage

Compressed air vehicles are not complex. They use a bottle of compressed air and some kind of compressed air motor (Figure 8-16). Neither is the concept new. Whilst Queen Victoria was still on the throne Whitehead used compressed air to drive torpedoes, and the French city of Nantes had compressed air trams.

Compressed air drive has several problems. It tends to be relatively inefficient. The energy to drive the car is transformed many times before it reaches the wheels: there are losses in the generation and transmission of the electricity, losses in the compressor and losses in the air motor.

There are significant thermal issues.

Compressing air heats it. Releasing the pressure cools it. In extreme cases valves may ice up. There may be further sacrifices in efficiency.

A high pressure bottle is also a liability in an accident. In July 2008 an oxygen bottle in the hold of an airliner ruptured, blowing a hole in the skin of the aircraft and projecting the bottle itself up through the floor of the cabin. Fortunately no one was injured.

Finally, even with very high system pressures, the specific energy (amount of energy stored per unit of weight) is relatively low: anecdotally, no better than lead-acid batteries.

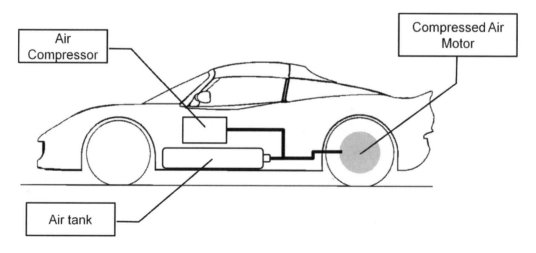

*Figure 8-16 Compressed air energy storage*

# Superconducting Magnetic Energy Storage

Superconducting Magnetic Energy Storage (SMES) makes use of two phenomena

- The field generated by an electromagnet stores energy which can be recovered as electrical power

- The electrical resistance of certain metals drops to zero at very low temperatures

Fundamentally an SMES is a coil of superconducting metal in a jacket of liquid helium. The SMES is "charged" by passing a current through the coil. As the magnetic field builds up, energy is absorbed. At a steady state, almost no power is required to maintain current flow. If a load is placed across the coil, the decaying magnetic field drives power through it.

Fixed SMES systems are in use in industrial environments for backup and power regulation. None are currently available for mobile use and the very low temperatures required adds a great deal of cost and complexity. Efficiency is however said to be very high and they are capable of producing very high power instantaneously.

# Inductive charging

Inductive charging could conceivably be used to recharge an EV on the move thus providing infinite range. It is a way of transferring power to an EV (or indeed any battery powered device) without a physical connection. This sounds like magic, but isn't. If you have read earlier chapters you will be very familiar with the idea that a voltage is induced in a wire that is placed in a moving magnetic field. Indeed this is the principle of a conventional transformer: an AC current in the primary windings creates a moving magnetic field in the transformer core. The moving magnetic field induces a voltage in the secondary windings (Figure 8-17). A transformer of this kind is most efficient with a continuous iron core, but would still function (albeit with reduced efficiency) with an air gap (Figure 8-18).

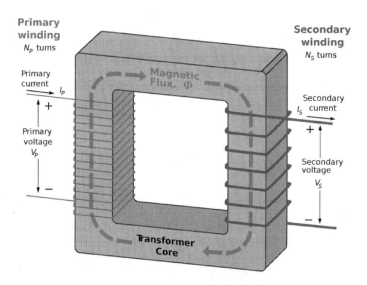

*Figure 8-17 Transformer (Illustration from Daniels, A 1976. Introduction to Electrical Machines. Macmillan Publishers. ISBN 0-333-19627-9, released under the GNU Licence.)*

Now imagine the transformer in Figure 8-18 rotated through 90° with the secondary windings mounted in a vehicle and the primary windings embedded in the road (Figure 8-19). This would allow a crude form of inductive charging.

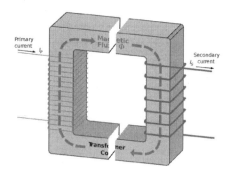

*Figure 8-18 Transformer with air gap*

A real-world installation would use a different mechanical design, but the same electromagnetic principle. Inductive chargers are readily available for consumer electronics such as games console remotes, and there have been some forays into inductive charging for EVs.

A very interesting longer-term possibility is inductive charging on the move. If roads (or sections of roads such as one lane in multi-lane motorways) were modified as inductive chargers, then suitably-equipped EVs could recharge whilst driving down the road. The principle has been illustrated by the e-quickie project at Karlsruhe University of

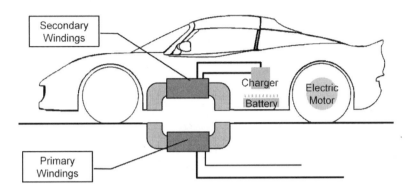

*Figure 8-19 Inductive charging schematic*

Applied Sciences (Figure 8-20).

The combination of a smaller battery and on-the-move charging is enticing. EVs could be made smaller, lighter, cheaper and even more convenient. The technology has its challenges however. The e-quickie, for all its success in showing what could be done, is a technology demonstrator; far smaller and lighter than any usable car and requiring the in-road charging infrastructure which does not yet exist outside of a test track. There would be a host of problems to solve from payment to handling snow. The potential infrastructure investment is intimidating. The use of the vehicle-to-grid ideas discussed in the next chapter might be problematic.

*Figure 8-20 E-quickie. Photo courtesy of Hochschule Karlsruhe – University of Applied Sciences. Photographer: Uwe Krebs (http://www.hs-karlsruhe.de/en/home.html)*

# Overhead power

*Figure 8-21 Toronto Tram from the 1920s. Note overhead power cables. Photo courtesy of David Arthur (licensed under Creative Commons Attribution-Share Alike 3.0)*

The transmission of power to a moving vehicle from overhead wires (or in the case of the London underground, a third rail) has been widely used for trains and trams for over a century. (Figure 8-21)

Overhead power cables are fairly easy to organise with trams or trains that run on fixed rails. Overhead wires have however also been used with conventional road vehicles. Buses of this kind are normally referred to as "trolley buses" and have been widely used around the world (Figure 8-22).

The superior low-rpm torque of an electric motor makes trolley buses attractive in hilly areas (note the twin rear axles in the Newcastle trolley bus in Figure 8-22). For example, San Francisco Municipal Transportation Authority claims to have the largest fleet of trolley buses in the USA[19]. Many modern trolley buses are a form of hybrid with batteries or a diesel engine providing power when not connected to the overhead wires.

---

[19] http://www.sfmta.com/cms/mfleet/trolley.htm

*Figure 8-22 Trolley bus of the kind used in Newcastle-on-Tyne in the 1960s. Photo courtesy of Steve Kemp (licensed under the Creative Commons Attribution-Share Alike 3.0)*

Even as an adjunct to battery power, it is difficult to see the same principle being applied to private cars. Quite apart from the very long pick-up poles required with a smaller vehicle, the sheer number of cars on the road, the problem of overtaking and numerous other complications would make it difficult to implement.

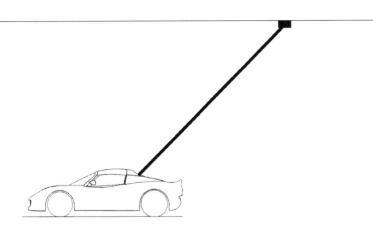

*Figure 8-23 Trolleycar anyone?*

# In Conclusion

Fuel cells, flywheels, hybrids, ultra-capacitors, compressed gas, superconducting magnetic energy storage, inductive charging on the move and overhead power: someone, somewhere has demonstrated every one of these power storage methods. Some are in use in real-world applications already.

The concept of inductive charging on the move is perhaps one of the strongest contenders for future adoption. Existing EVs would need little modification to make use of inductive charging facilities, and would not absolutely require such facilities to be viable.

It is however difficult to see any of these technologies supplanting the pure battery EV altogether. Most either have substantially lower specific energy than Lithium-Ion cells or high cost, or both.

# 9: Why Bother with Battery Power?

Previous chapters have dealt with the technological issues surrounding EVs. In this final chapter we address an issue which for many people comes first logically. What are the motivations for making the change from our IC to EV? We have been driving IC-engined cars for nigh on 100 years. Over that time they have been transformed from nasty, noisy, unreliable devices to a level of smoothness, reliability and efficiency that is remarkable given their complexity. So why bother with EVs?

Two reasons are often given –

- "peak oil" (i.e. it's running out)

- climate change

Unfortunately both of these are contentious issues: some hold that peak oil is exaggerated and some that mankind's activities have little to do with climate change. If these were the only benefits of EVs then there would be some argument about whether the change was worth making. There are however many other powerful motives for adopting electric drive. We cover some of these, as well as addressing the "peak oil" and climate change issues.

## Urban air pollution

Anyone who has been through a busy road tunnel knows what other people's exhaust fumes smell like. Those of us who live in big cities (i.e. most of us) spend much of our lives breathing in dilute vehicle exhaust (Figure 9-1). We would not expect this to be good for us. There is evidence that diesel emissions in particular have a number of harmful effects. For example, the American Journal of Epidemiology reported an apparent link between exposure to pollutants typical of diesel engines and

*Figure 9-1 Exhaust fumes, especially diesel particulates, are bad for human health. Photo courtesy of Marius Zaharia (http://www.flickr.com/photos/mordax/with/3183266304)*

impaired brain development in children[20]. The New England Journal of Medicine reported a study[21] in which 60 volunteers, all asthmatics, walked down Oxford Street (diesel buses and taxis only), then Hyde Park (traffic free). There was a major difference in the level of both particulates and the aggravation of asthmatic symptoms.

There are many other studies suggesting links between diesel emissions and asthma, cancer, impaired brain development and other issues.

Newer diesels are better than older ones in terms of air pollution when correctly adjusted. Allowable levels of particulates and nitrogen oxides (the main baddies) are being steadily lowered for new cars in both the USA and Europe. The manufacturers can however only achieve the new standards by adding expensive particulate filters and/or catalytic converters. And it isn't just the basic equipment either. Extra electronics are also needed to monitor emissions to ensure that they don't drift outside the limits.

From an engineering viewpoint, making a dangerous mess, then cleaning it up on the

[20] November 2007
http://aje.oxfordjournals.org/cgi/content/abstract/kWm308v1

[21] December 2008
https://content.nejm.org/cgi/content/full/357/23/2348?ck=nck

186

way out of the exhaust pipe seems very inelegant. By contrast an EV charged from a renewable source (such as wind or hydro) has zero emissions, and even if electricity from a coal-fired power station is used, the indirect emissions are physically separated from population centres and are in any case more readily controlled.

# Noise pollution

A number of studies have suggested a link between traffic noise and hearing loss. For example an Indian study published in 2009 found very substantial hearing loss in shopkeepers working near a busy road[22].

Hearing loss is bad enough, but rather more seriously, a study by the Karolinska Institutet in Sweden[23] reported in the British Medical Journal suggested that people living in areas of high traffic noise have a greater risk of myocardial infarction (heart attack).

Also, as long ago as 2001 an Austrian study reported that "...children living in relatively noisy neighbourhoods had raised blood pressure, heart rates and levels of stress hormones..." This was at sound levels far below the 85 decibels or so that are currently considered to be damaging to hearing when frequently repeated.

So, conventional vehicles are implicated in hearing loss, raised incidence of heart attacks, cancer, asthmatic problems, impaired brain development and increased levels of stress in children. That ought to be enough to encourage us to look for alternatives.

---

22

http://www.informaworld.com/smpp/content~content=a906346013~db=all~jumptype=rss

23

http://ki.se/ki/jsp/polopoly.jsp?l=en&d=130&a=69349&newsdep=130 and

http://oem.bmj.com/cgi/content/abstract/64/2/122

# Peak oil, or the Grandkid's Aviation Fuel

So EVs have major benefits for our health and well being and on the health and well being of our children, but there is another issue affecting our children: the accelerating rate that we are getting through the world's oil (Figure 9-2).

When Stonehenge was built, what we now know as the North Sea Oil and Gas field already existed pretty much in the form in which it was discovered 4,000 years or so later. For all their exploits, the builders of

Stonehenge had no idea that it was there, and could not in any case have extracted any of it.

It was still there nearly 2000 years later when Julius Caesar landed in what is now Kent. The Vikings crossed over it quite regularly 800 years after Caesar, again with no clue that it existed. Another 1000 years on and Admirals Jellicoe and Beatty sailed across the oil fields in their coal – powered grand fleet to the Battle of Jutland, little

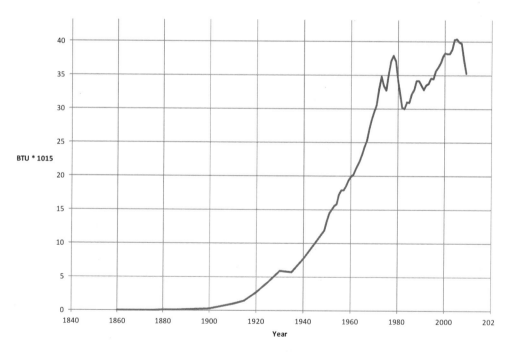

*Figure 9-2 US Oil Usage 1860 - 2009 (Source - US Energy Information Administration)*

knowing that beneath their keels was the stuff that would power warships of the next great war. 50 years later the long sleep ended: North Sea oil was at long last detected. Within another 50 years we had dug most of it up and burned it.

50 years in historical terms is the blink of an eye. North Sea oil is a microcosm of our whole petroleum-based culture: it's a short-term oddity, a developmental dead-end. Whether you believe that oil was laid down by the decay of animals in the Triassic or created along with the rest of the earth's crust a few thousand years ago, most of us believe that once it's gone, it's gone. You don't have to be a tree-hugger to think it unlikely that our grandchildren will still be driving cars powered by petroleum products on the scale that we are. There simply isn't enough of the stuff left in the ground.

If we were responsible we would use the oil we have only for activities where there is no viable alternative, such as aviation. Whilst we know several ways to power cars, commercial aircraft have always used fuels derived from petroleum. We know no other scalable way to fly people across the Atlantic. So if we really cared about our children and grand children, we would reserve oil for things like aviation where there are no alternatives. As things stand today, though, aviation accounts for just 12% of the fuel consumption of the entire transport sector, compared with 80% dedicated to road transport. We are using most of the little oil we have left in road transport where, with a little effort, we could use something else.

EVs need energy but they don't need oil. You can power an EV from hydro-electric power, nuclear power stations, wind energy, tidal or wave energy, even (as costs come down) from solar panels on the roof of your house. Burning precious oil to drive cars is vandalism. We may stand accused by our grandchildren of burning the furniture because we couldn't be bothered to go out and bring in some logs.

# Biofuels

First generation biofuels (broadly liquid fuels derived from foodstuffs such as corn) were described by Jean Ziegler of the UN[24] as a "crime against humanity". According to his report "...232kg of corn is needed to make 50 litres of bio-ethanol.... a child could live on that amount of corn for a year".

A World Bank report leaked by the *Guardian* newspaper opened with the startling statement: "The World Bank's index of food prices increased 140 percent from January 2002 to February 2008. This increase was caused by a confluence of factors but the most important *was the large increase in biofuels production* in the US and EU...." [my italics]

Three years later, the effect of first generation biofuel production on world food prices remains a concern. It has also been realised, somewhat belatedly, that the production of the crops that go into these fuels itself consumes energy (and generates so-called "Green House Gases") in amounts that largely negate the supposed benefits. Lastly there is an issue of scalability. It takes a huge area of land to grow enough fuel to drive a car for a year, for much the same reason that it takes a large area of woodland to grow enough wood to keep a house warm using wood-burning stoves: if the US turned over its entire grain harvest to ethanol production, it would still have to find over 80% of its transport fuel from elsewhere.

Second generation biofuels are produced from non-crop sources such as stalks, stems and other waste left over from food production. This reduces the impact on food prices but there are even greater concerns over scalability: think of the amount of food waste produced in your household, then think of three lanes of bumper-to-bumper traffic on the M25 on a Friday night. The numbers don't begin to work. Furthermore there is concern over the energy costs in transporting vegetable waste and by-products to biofuel production plants.

---

[24]

http://www.un.org/apps/news/story.asp?NewsID =26478&Cr=food&Cr1=ziegler

Another possible source of biofuels is algae. This is sometimes referred to as "2nd Generation" and sometimes as "3rd Generation". Algae-based biofuels promise to address the scalability issue to some degree (algae are far more efficient than conventional agriculture at producing biofuels). However there are concerns over cost and the time this technology will take to reach commercial viability.

The widespread adoption of electric vehicles makes this problem go away. If we don't need liquid fuels at all we don't need to make biofuel. If we really want to power transport using biomass we can revert to a technology James Watt would have recognised: generate electricity from steam produced by burning wood (Figure 9-3).

*Figure 9-3 The original bio fuel powerplant. Photo courtesy of James Howe Photography (http://jameshowephotography.com)*

# Climate Change

At some point in books and articles on EVs and hybrids, it is normal to go into their impact on climate change. This can be off-putting for the significant minority who are not convinced that climate change is man-made (or "anthropogenic" - i.e. caused by our activities). I do not propose to take sides in this debate as EVs are a win-win. If anthropogenic climate change is important for you then anyone who drives an EV is doing you a favour, regardless of why they are doing it. If you are keen on EVs for other reasons (such as urban air and noise pollution as described earlier), then those who are motivated by climate change are also supporting your agenda.

Briefly (for anyone who has been on silent retreat for the last 20 years) the issue is that average global temperatures have risen over the last 100 years or so, albeit with a dip between the 1940's and the 1980's and (according to the Met Office Hadley Centre) a plateau since about 2000 (Figure 9-4)

The currently-dominant hypothesis is that this is due to the warming effect of the release by mankind into the atmosphere of "green house gases" (GHGs) – primarily carbon dioxide, methane, nitrous oxides and refrigerants, and that if we allow this to continue we (or our children/grandchildren) face a strong probability of dire

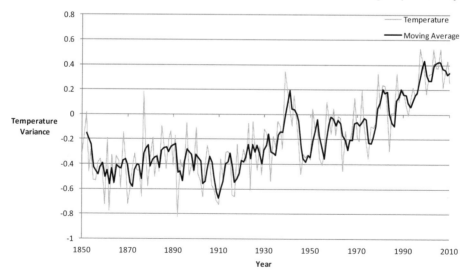

*Figure 9-4 Mean Global Temperatures (HADCRUT3 data)*

consequences (primarily floods, crop failures, rising sea levels and some of the hotter places on earth becoming virtually uninhabitable).

On this model, the over-riding concern is to minimise or eliminate the release of carbon dioxide into the atmosphere. For road transport the important figure is the grams of carbon dioxide produced per mile (or km) travelled.

Battery powered vehicles offer a substantial improvement to this grams-per-mile figure, but not a complete solution.

# Energy Efficiency and GHGs

It is often said that EVs have neither markedly better well-to-wheel efficiency nor better GHG numbers than the best of conventional IC engines: that they merely shift the fuel consumption and GHG production from the car to the power station. There are a number of papers by reputable organisations that purport to demonstrate this. Numerous media stories carry a similar message.

None that I have read bear close examination. They are all based on plausible, but ultimately misleading, assumptions. Here are some of them:

## Ignore the power consumed extracting and refining petrol

Extracting and refining petrol itself consumes power (Figure 9-5)

Extraction efficiency ranges from 78% to 92%[25]. Refining efficiency is also around 90% [26] . The two added together are therefore about 80%. Energy content of a US gallon of petrol is around 32 to 34 kW-hrs[27]. Thus the energy used in extracting and refining a US gallon of petrol might be around 6 to 8 kW-hr. A typical EV achieves

[25] "Fact Sheet: Energy Efficiency of Strategic Unconventional Resources" [US] DOE Office of Petroleum Reserves

[26] See "Estimation of Energy Efficiencies of U.S. Petroleum Refineries" Michael Wang, Center for Transportation Research, Argonne National Laboratory, March 2008 and "Updated Estimation of Energy Efficiencies of U.S. Petroleum Refineries" Ignasi Palou-Rivera and Michael Wang July 2010

[27] The US EPA estimate that the energy content of petrol varies between 108,000 and 117,000 BTU http://www.epa.gov/oms/rfgecon.htm#1

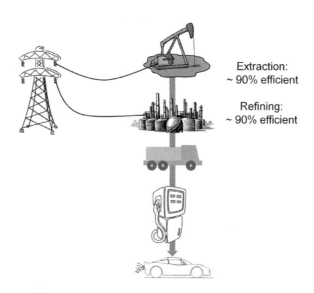

Extraction:
~ 90% efficient

Refining:
~ 90% efficient

*Figure 9-5 Petrol Extraction and Refining losses*

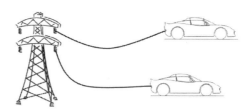

*Figure 9-6 Some EVs can travel further on the energy used to extract and refine petrol than an IC engined car can drive on the petrol itself*

about 4 to 5 miles per kW-hr so if these numbers are correct, an EV could travel around 24 - 30 miles on the power required merely to *extract and refine* a US gallon of petrol. (Figure 9-6).

The average in-use fuel consumption for passenger cars in the US is only 21.5 m.p.g.[28]. It therefore seems likely that some EV drivers go further on the power needed to extract and refine a gallon of petrol than many IC engined cars go on the petrol itself: although to be fair, the typical EV is smaller and lighter than the average US passenger car.

_____

[28]

http://www.epa.gov/oms/consumer/f00013.htm

194

## Use official fuel consumption figures for IC engined cars

Efficiency comparisons invariably use the official fuel consumption figures published by national bodies (such as the Vehicle Certification Authority in the UK). Real world average fuel economy figures for IC-engined cars are however nowhere near these official figures (often quoted in the small print of advertisements). "Highway" fuel economy figures are measured for a car in good condition, tuned and adjusted to perfection, operated by a skilled driver in mild weather, with few ancillaries running and with the engine fully warmed up.

There is much anecdotal evidence that even on long trips with a thoroughly-warm engine few people get close to the official figures. At the other end of the scale, the first mile or two of a school run or a trip down to the shops on a cold winter morning consume vastly more fuel than you would expect from the official figures[29]. This is

---

[29] In the UK, the AA noted "...Even at relatively mild winter temperatures, the fuel consumption of a cold car leaving its driveway is 40 per cent higher than normal..."
(http://www.theaa.com/motoring_advice/news/aa-fuel-for-thought-increased-cost-of-winter-motoring.html)

made worse by the power consumed by rear screen heaters, seat heaters, headlights, windshield wipers etc – all of which would be switched off on the official tests.

And recall that most cars are driven fewer than 40 miles a day so a high proportion of real-world driving is done with a cold engine. This has a major impact on real world fleet-wide fuel consumption.

## Ignore traffic jams

Another issue is nose-to-tail stop-start traffic. This has a disastrous effect on the per-mile emissions of IC engined vehicles, as anyone who has endured a traffic jam in a tunnel can testify. It has little effect on the per-mile energy consumption and no effect on direct emissions of EVs.

## Assume that the IC car is always correctly tuned

Not all cars are operating at their optimum all the time. According to figures released by VOSA[30] over 1.7 million cars in the UK failed their MOT tests in 2009/10 on the "fuel and exhaust" area. IC engined cars

---

[30]

http://www.dft.gov.uk/vosa/repository/10%2018 0a1.pdf

are, as has been noted, complex devices and the partial failure of obscure components such as lambda (oxygen) sensors can have a major impact on both fuel consumption and emissions. It is reasonable to assume that a high proportion of these MOT failures will have been exhibiting higher-than expected fuel consumption and/or emissions. This will push up the real-world national fleet average for IC engined cars by an indeterminate but possibly significant amount. EVs are vastly simpler mechanically and do not suffer from this

kind of generalised efficiency degradation.

## Ignore fuel distribution

There are energy costs associated with maintaining a network of petrol stations and keeping them all supplied with fuel. These include the energy associated with distributing thousands of tons of fuel from refinery to filling station, the additional miles driven by customers of those filling stations and the lighting and heating associated with operating those filling stations (Figure 9-7). It is difficult to quantify these losses exactly, but here are some order-of-magnitude estimates:

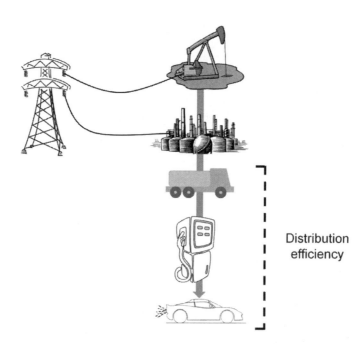

*Figure 9-7  Distribution Efficiency*

One UK tanker operator claims to cover 6 million miles a year delivering 2000 million litres of fuel[31]. If we assume 10 m.p.g. (generous to the tanker), then road tankers in this fleet burn around 1.3% of the fuel they deliver.

A 5 mile diversion by car drivers every 300 miles to get to a filling station would represent about 1.6% efficiency reduction.

Forecourt lighting is a major energy cost for filling stations. A typical small petrol station might sell 38,000 litres/week or about 2 million litres/year. The calorific value of 2 million litres of petrol is about 70 million megajoules or (very approximately) 20,000 megawatt-hours. One study suggested an energy consumption of 40 megawatt-hours per year just on forecourt lighting – that would represent another 0.2% of the energy in the fuel sold: just to light the forecourt.

These are relatively small numbers but they add up: just these factors alone would represent fuel consumption equivalent to over 3% of the energy content of the fuel sold.

---

[31] http://www.fleetdirectory.co.uk/fleet-news/index.php/2010/09/30/suckling-transport-improve-fleet-efficiency-while-recording-huge-falls-in-the-severity-and-frequency-of-crashes

## Compare the best IC car with the worst power stations

Older power stations are less efficient, and coal-fired power stations produce a great deal of $CO_2$. For a variety of reasons, our current electricity generating infrastructure contains a lot of coal fired power stations which can be made to look very bad from a GHG standpoint[32]. The comparison is drawn between *idealised* figures from the best modern diesel (ignoring the beat-up old rust-buckets and the V8 petrol 4x4s) on the one hand, and *actual* performance of the current generating infrastructure with present-day fuel mix on the other (Figure 9-8).

---

[32] However, from a fuel-independence point of view, the replacement of oil with coal is a huge win: whilst our coal production is low, we have enormous reserves of the stuff left in the ground which could be exploited. Also we expect carbon capture to become possible with fixed power stations but not with cars.

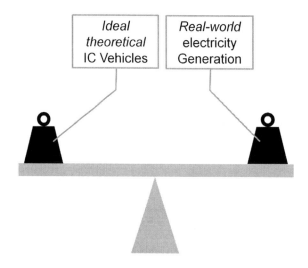

*Figure 9-8 An unfair comparison*

Back in the real world, the lifetime efficiency of a typical conventional car doing large numbers of short journeys with a cold engine in heavy traffic will be a lot less than the theoretical efficiency figure. For these and other reasons battery vehicles can be expected to bring about a significant real-world reduction in total energy use and emission of pollutants.

# The impact on electricity generation

It is sometimes argued that a large fleet of battery vehicles would swamp the electricity generating infrastructure of many countries including the UK. In fact the opposite is likely; a large fleet of EVs could easily become an integral and indispensable part of that infrastructure. With the addition of some fairly simple electronics, a fleet of EVs would provide a unique opportunity to set up a scalable mechanism for buffering or smoothing electrical power.

Even in the worst case, the numbers suggest that EVs would not in fact require major additional generating capacity. The power consumption of EVs would be a small percentage of current peak demand from commercial and domestic customers. Department for Transport statistics[33] show

---

[33] See:

http://www.dft.gov.uk/pgr/statistics/datatablespublications/vehicles/

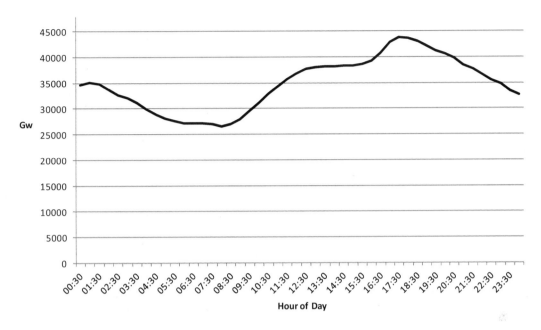

*Figure 9-9 UK Power Demand on 5th January 2011*

an average UK annual mileage for cars of about 8,800 each. That is about 24 miles/day per car. If we assume between 4 and 5 miles per kW-hour, the average daily energy required by a battery vehicle at 24 miles per day is 5 - 6 kW-hours.

Even if we make the severe assumption that recharging is all done during the day (i.e. we spread the load over a 12 hour period rather than 24 hours), the power required is on average about 0.5 kW (6 kW-hours divided by 12 hours) per car or 500,000 kW for 1 million EVs. The UK generating capacity is around 80 GW i.e.

80,000,000 kW [34] - so the first million battery-powered vehicles would represent only about 0.6% - 0.7% of current grid capacity. That is so small as to be virtually insignificant.

Even if all 26 million private cars on Britain's roads were eventually replaced by battery-powered ones, we would only be talking about a 16% - 17% uplift in power requirements.

---

[34] 2008 figures – see for example http://www.nationalgrid.com/uk/library/docume nts/sys05/default.asp?action=mnch3_11.htm&N ode=SYS&Snode=3_11&Exp=Y#Power_Statio n_Capacities

There is another major factor. The experience of early adopters is that most charging of EVs takes place, not during the day at public charging stations, but at night. Electrical demand falls off precipitously in the late evening (Figure 9-9). If most EV charging occurred in this period substantial EV-related demand could be added without *any* significant increase in peak demand.

So much for the power requirements of EVs. Now for the benefits for power generation, which have the potential to be huge. The big problem for electricity generation is that (apart from expensive pumped storage) electricity cannot be stored in a scalable fashion. It must be produced the instant that it is needed. This means that some generating capacity must be kept in reserve, ready to meet the huge fluctuations in demand that can occur, such as when (for example) everyone switches on their electric kettle at half time in a major nationally-televised football match (see Figure 9-10). Some of this is so-called "spinning reserve" where a certain amount of generating capacity is operating at partial load (or even zero load). This has a potential impact on efficiency. Sometimes the reverse happens: an event triggers a sudden drop in demand (see Figure 9-11); this again reduces efficiency.

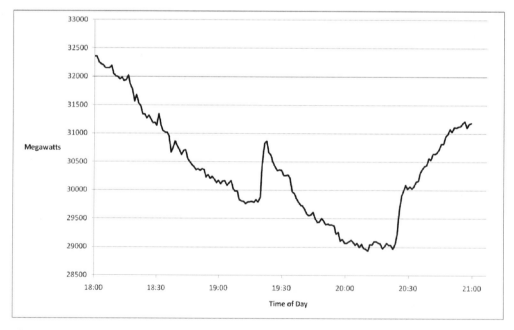

*Figure 9-10 Half time surge 12 June 2010 - Data for Figure 9-10 and Figure 9-11 courtesy Energy Requirements, Network Operations, National Grid*

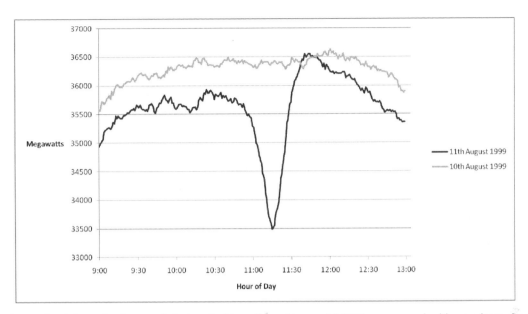

*Figure 9-11 Drop in demand during Solar eclipse (August 1999) compared with previous day*

Most cars spend most of the time parked. The batteries in a fleet of parked EVs could (assuming they were plugged in and had the right electronics) be used as a huge buffer for smoothing out such peaks and troughs. This is sometimes referred to as "vehicle-to-grid" or "V2G".

Order-of-magnitude calculations suggest that this could be significant. A typical 100 mile range EV might have a battery capacity around 25 kW-hours. If 40% of cars on the road were EVs of that class, that would be a little over 10 Million such cars. They would between them be capable of storing 25,000 Megawatt-hours. If (for example) contracts existed between the

power companies and the EV owners that up to 10% of the battery capacity could be "borrowed" back as required, then up to 2.5 GW hours would be available. This would be more than enough to handle half-time surges.

If 80% of Britain's cars were EVs, and if battery advances meant that most cars had 75 kW-hour packs then 15 GW hours could in theory be made available. To put that in perspective, that would be sufficient to run the entire country *with every power station shut down* for between 15 and 30 minutes depending upon the time of day. The picture becomes even more enticing when you start to factor in renewable energy sources such as wind, solar, tidal etc. The

201

sad reality is that we could cover the country in windmills without having any effect at all on the conventional generating capacity we need. Sometimes the wind doesn't blow, and Murphy's Law suggests that peak power demand will sometimes occur on calm winter evenings when output from wind farms and solar arrays is zero. Even tidal power has four short periods a day around slack water when power output will drop to zero.

A national fleet of EVs equipped for V2G is the perfect adjunct to renewable energy sources such as wind, solar and tidal power. During periods of slack demand and high power availability, the excess energy can be used to charge vehicle batteries. When the wind dies down and demand goes up, power can be borrowed back from those same vehicle batteries.

# The negatives

Those are all the positive benefits of EVs and they are an impressive list. So what are the negatives? Surely battery powered vehicles have their drawbacks?

Well yes – their key deficiency is the limited power that can be stored in today's batteries. Pure electric vehicles will in the near future have limited range, limited carrying capacity and a high price.

Hybrids have little or no range limitation, but suffer from the implicit drawback of having two of everything: an electric motor and an IC engine, a fuel tank and a battery, a fuel injection system and a controller, a battery charger and a fuel filler cap, a fuel gauge and a Depth of Discharge indicator. This

can only make them heavier and more expensive than either a conventional car or a pure EV. This is not unknown however. Modern sailing vessels inevitably have engines as well as sails. Aircraft have wheel brakes and steering as well as flight controls. Most submarines use diesel power on the surface and battery power submerged.

It is also worth considering the impact on our economic infrastructure. The currently-dominant vehicle manufacturers have been slowly refining IC drivetrains for a hundred years. If EVs take over from IC engined vehicles then all that experience, all those patents, all that commercial advantage, become worthless. The risk for the big motor manufacturers is that in the upheaval caused by EVs they will lose out to newcomers in the way that the dominant manufacturers of mechanical watches lost out to new electronic watch brands in the 1970's. The potential impact on western societies in particular could be drastic.

EVs need far less maintenance than IC engined vehicles. A major move to such vehicles would put half the world's car maintenance facilities out of business and make half the world's motor mechanics unemployed. For governments, a big risk is the loss of tax revenue. Road transport fuel is easy to tax separately because it only has one use, and it is a major taxation revenue stream. Electricity is a core component of our commercial and domestic energy infrastructure and taxing electricity used in road transport would be next to impossible.

The commercial interests of the oil companies and oil – dependent economies are too obvious to elaborate. No one in commercial life likes to see a major threat to one of their major revenue streams. There are many other areas of commercial life from oil refining to commercial vehicle production that would be affected to some degree.

# Resource consumption by EVs

It is sometimes said (or implied by innuendo) that the widespread adoption of electric vehicles would merely shift the problem of resource consumption from one commodity (oil) to another (raw materials for batteries).

There is some truth and some error in this. Broadly speaking there is not enough lead in the world to support universal adoption of EVs powered by lead-acid batteries. Nickel and cadmium stocks are likewise limited, meaning that nickel cadmium or nickel metal hydride batteries are probably never going to dominate if there is a wholesale move to EVs. Cobalt for lithium cobalt batteries is also in short supply.

There is however enough lithium. One 1999 paper[35] concluded that a couple of lithium ion chemistries looked practicable for 2 billion EVs or more.

---

[35] "Large-scale Electric-vehicle Battery. Systems: Long-term Metal Resource. Constraints" by B. A. Andersson and I. Råde..

Electric vehicles have the potential to make our urban areas quieter, pleasanter and healthier places. Widespread adoption of EVs would reduce our wanton consumption of a precious resource that was given to all generations, not just ours. EVs do not steal food from the dinner tables of the world's poorest people

There is however another good reasons to adopt them. They are more fun. People who drive EVs mostly wouldn't go back to IC engines if you paid them. EVs are in a different league when it comes to low end torque, and it is low end torque which makes a car feel responsive on the road. Travelling in an open-topped electric sports car on a fine evening is a remarkable experience: surges of power accompanied by nothing more than a bit more wind noise.

# A final parable

Let me finish with a cautionary tale from a different era and a different domain. The piston engined Hawker Tempest and the jet engined Gloster Meteor entered RAF (Royal Air Force) service in 1944 within a few months of each other. The Hawker Tempest was one of the last of an illustrious line of piston engined fighters powered by huge engines honed by years of war. The Meteor was the first of the new breed of jet engined fighters. With a top speed a little over 430 m.p.h, the piston-engined Tempest was around 20 m.p.h. *faster* than the contemporary jet engined Mk I Meteor (Figure 9-12).

Anyone looking at those numbers would have wondered why the RAF bothered with

*Figure 9-12 Piston Engined Tempest (left) and jet-engined Meteor (right).*
*The Tempest was 20 m.p.h. faster than the Mk 1 Meteor (Tempest photo courtesy of*
*Gerald Trevor Roberts, Meteor photo courtesy of Nicholas Foss)*

jet engined fighters. 18 months later they would have had their answer: a Meteor took the world air speed record at over 600 m.p.h: 80 m.p.h. faster than the *current* air speed record for piston engined aircraft.

Any new technology tends to start off at a disadvantage when compared with the incumbent technology. It happened with the jet engine and it happened with digital watches: and who can now credit that the earliest pocket calculators weighed half a kg and cost the equivalent of £1000 in today's money?

When it comes to EVs we have not seen anything yet. We are still at the Mk1 Meteor stage of development. The first jet engines were bulky, temperamental, unreliable and had short service lives. Current EVs have poor range. Batteries are expensive, bulky and need careful management. All these problems will go away just as they went away for jet fighters in the first few years of the jet age.

If EVs become mainstream they will make piston-engined cars obsolete just as jet fighters rendered piston-engined fighters obsolete. Piston-engined cars won't disappear altogether – enthusiasts will still run them as a weekend hobby, just as enthusiasts still fly old Mustangs, Spitfires, Bearcats and Tempests: but no one will use them for daily transport.

# 10: Selected Bibliography

I've mentioned a number of papers in the text and included data about them as footnotes. Here are a number of books for further reading. I've included both the good stuff and also some that looked interesting but which I cannot recommend. Apologies to any authors who feel misrepresented. Works are listed in alphabetical order by first author. The comments are all mine, although they have in most cases been previously published elsewhere.

# Good

## Owning an Electric Car: Discover the Practicalities of Owning and Using Electric Cars- for Business or Leisure

by: Boxwell, Michael      Date: 2010      ISBN: 1907215107

Published by: Code Green Publishing, Paperback, 200 pages

This book deals with owning and operating an EV. The author speaks with authority on a topic he clearly knows well and has researched carefully. He writes lucidly and objectively, marshalling facts and letting the reader draw his or her own conclusions. What more could one ask?

The book is not very demanding mathematically: indeed it does not go into much detail about different EV technologies, but that is not what it is about. There are other books for that. If (like me) you react badly to the word "green", but care about your grandchildren's energy supplies, the section on Electric Cars and the environment is worth the price of the book even if you are a petrol head.

## Bottled Lightning: Superbatteries, Electric Cars, and the New Lithium Economy

by: Fletcher, Seth       Date: 2011       ISBN: 809030535

Published by: Hill and Wang, Hardcover, 272 pages

Excellent. The author is a science reporter and this book is both an easy read and also informative and well researched (it was recommended to me by an academic, a Fellow of the Royal Society which says it all really). It covers the history of Lithium cells and puts flesh on the dry bones of Lithium supply. A nice choice of title too!

## Electric Motors & Drives

by: Hughes, Austin       Date: 2005       ISBN: 750647183

Published by: Newnes, Edition: 3nd, Paperback, 384 pages

This book is brilliant. It isn't stylish, it's not lavishly illustrated, It isn't even a riveting read. It is, however, plain old fashioned lucid. Hughes was an academic, and if he lectured as well as he writes, his students understand electric motors.

He also speaks with authority. The diagrams are small, mostly simple black-and-white drawings, but they fit in with the text, It isn't for everyone though. You'll want at least A level physics to understand it and (despite his protestations) it is fairly mathematical.

By far the best book I've read in this subject area for a while.

## Modern electric, hybrid electric, and fuel cell vehicles: fundamentals, theory, and design

by: Ehsani, Mehrdad     Date: 2004       ISBN: 849331544

Published by: Boca Raton, FL : CRC Press

This book is just what it says on the tin - fundamentals, theory and design. The maths is extensive but not difficult: roughly O-level/GCSE rather than university level. It is quite well laid out, explained and written. It covers (among other things) transmission topologies, motors, batteries, controllers and fuel cells. The first section (essentially generalised tree-hugging) is a bit of an aberration and may irritate those who are uncertain about anthropogenic global warming

## Witness to Grace

by: John B. Goodenough       Date: 2008      ISBN: 1604747676

This is the autobiography of one of the most important (and maybe most under-rated) scientists of the late 20th/early 21st century. John Goodenough invented the lithium-cobalt battery, and a member of his team invented the Lithium iron phosphate battery. He was also one of the key figures in the development of solid-state memory, without which we would have no PCs He was for many years a professor at Oxford and in his late eighties was still professor of mechanical engineering and materials science at the University of Texas at Austin.

The book itself is fairly short: an eclectic description of a personal, scientific and spiritual pilgrimage. Reflections on biblical verses rub shoulders with passages about itinerant electrons and transition metal oxides. Fortunately you don't have to understand either to profit from the book. There are vignettes, brief but fascinating, about co-operation with scientists in many countries including the cold war Soviet bloc.

# Middling

## Battery Management Systems for Large Lithium Ion Battery Packs

by: Andrea, Davide      Date: 2010      ISBN: 1608071049

Published by: Artech House (2010), Edition: 1, Hardcover, 300 pages

There is a school of thought that BMS systems do more harm than good. This book is an authoritative presentation of the "pro" camp from someone who builds BMS systems for a living, and is useful for anyone wants to understand both sides of the argument.

The book deals with a complex subject in a largely non-mathematical way; although at least GCSE (maybe even A level) physics would be needed. For sections of the book you also need some familiarity with the jargon of modern electronics - MOSFETS, ASICs, Multiplexers and the rest - but the book still has value if you skip those bits. To the author's credit he doesn't big up his own products, and uses plenty of examples of competitive devices. He also to his credit goes into some details of precautions to take when wiring up circuits, and give an example of a converted Prius, burned out for lack of a washer on a battery connection.

The book has in my view three weaknesses. Firstly the big question with Lithium Ion battery packs is the extent to which individual cells in a pack drift apart in capacity and state of charge. There is published test evidence of this with AGM (a form of valve regulated lead acid) batteries, but the author presents no hard documentary evidence to confirm that Lithium chemistry cells behave the same way. Secondly his arguments for top-balancing rather than bottom balancing are rather shaky. Thirdly, it is an expensive book.

## Build Your Own Electric Vehicle

by: Brant, Bob   Date: 1993      ISBN: 830642315

Published by: McGraw-Hill/TAB Electronics (1993), Edition: 1, Paperback, 310 pages

This review covers the 1993 edition. Seth Leitman's update is reviewed below

The book is mixed - good in parts, not so good in others. I found that the explanation of controllers very useful. The stuff about EVs and the environment was a bit tedious. The whole book naturally suffered from being a bit dated but much of it was still relevant (depressing how little real progress has been made in the last 19 years). Technically it is pitched around Physics A level.

## Electric Motors - Workshop Practice Series

by: Cox, Jim      Date: 2007     ISBN: 1854862464

Published by: Trans-Atlantic Publications, Inc., Edition: 2nd, Paperback, 134 pages

The good bits: the author writes with authority on a subject that he clearly knows well. The book seems well organised and is generally well written and illustrated clearly.

The not-so-good: the book is a little bit limited in its scope, with all the examples being small motors. You won't find the kind of motor used in the Tesla for example. I'd have liked to have seen a bit more on brushless DC motors. Also (a minor carp) some of the photographs were a bit dark and a bit small. Whether you find the book useful probably depends on what you are looking for and what else you have read. If you run a small machine shop, I guess it would be ideal. I'm glad I read it but I probably won't read it twice.

## Lightweight Electric/Hybrid Vehicle Design (Automotive Engineering)

by: Fenton, John          Date: 2001          ISBN: 750650923

Published by: Butterworth-Heinemann, Perfect Paperback, 320 pages

This is a highly academic work. If you don't have an engineering degree or equivalent, don't bother. Even at that, the structure is, frankly, muddled. The diagrams are poorly labelled and, one suspects, sometimes chosen for availability rather than to serve the flow of the narrative. Concepts and organisations pop up at random. For example PNGV is mentioned on page 61, but not explained to p98.

Most of the last 75 pages relate to vehicle structures and suspension for efficiency that apply to any vehicle; not just electric ones. Most of the references and recommendations for further reading date from the mid nineties or earlier. The only redeeming feature is that the authors clearly know their stuff

## Electric Motors and Control Techniques

by: Gottlieb, Irving          Date: 1994          ISBN: 70240124

Published by: McGraw-Hill/TAB Electronics, Edition: 2, Paperback, 304 pages

This book is in many ways a good effort. It is detailed and explores some unusual motor configurations. The first third is about various designs of electric motor and most of the rest is about controllers, with a shortish section at the end about electric vehicles and discussion of hyperflywheels, cold fusion and the like.

The main problem for me was that it was dated. There was no mention of IGBTs for example and in the (brief) discussion of Electric Vehicle batteries, no mention of Nickel metal-hydride batteries. Lithium batteries were only mentioned as little more than a research project.The other thing I struggled with was that the text was poor at explaining principles: a few paragraphs of description, then it would slap a circuit diagram in front of you.

## 21st Century Hybrid Car and Hybrid Electric Vehicle Technology Assessment, Report for the Department of Energy

by: Government, U.S.    Date: 2005          ISBN: 142200032X

Published by: Progressive Management, Ring-bound, 138 pages

This work is a cross between an article and an academic paper. It is quite readable. It's based wholly on a series of simulations using the ADVISOR package. One appendix is quite mathematical but the rest requires no more than A-level understanding to interpret graphs etc. Despite its date, the content is a little dated though. NiMh batteries are assumed. Also the baseline comparison conventional vehicles are US mid-sized petrol vehicles. Comparison with a European diesel would have yielded different results. They dismiss out of hand the idea of a series hybrid with an engine sized for average power requirements - which is close to what the PML Flightlink Mini has already demonstrated.

There is an interesting discussion of "grid-connected" (plug in) hybrids. They investigate a hybrid configuration with an IC engine sized to handle gradeability (20 minutes up 6.5% grade at 55mph), plus Electric Motor sized for acceleration. This is a timely reminder of the careful trade-offs in specifying engine and motor sizes.

## Propulsion systems for hybrid vehicles

by: Miller, John M.      Date: 2004      ISBN: 863413366

Published by: London : Institution of Electrical Engineers.

This is a very detailed and quite highly mathematical coverage of hybrid architectures, motors, batteries, controllers etc. The strength of this book is its authoritative and carefully-researched content and good references. Its weakness is that it does not explain many of the concepts and terms that it uses. This is a book which will send the beginner to sleep but should deepen the knowledge of readers who are already familiar with the content.

In summary, a book for experts only

## The complete idiot's guide to hybrid and alternative fuel vehicles

by: Nerad, Jack R.      Date: 2007      ISBN: 9781592576357

Published by: New York, N.Y. : Alpha Books, c2007.

This is a survey of then-current options for alternative propulsion cars in the US. It is non-technical (it deals more with behaviour than construction details). The content is quite well-balanced: for example it addresses "green" issues but doesn't assume that the reader is a tree-hugger. Very properly in my view, the author handles man-made global warming as theory rather than fact without requiring the reader to adopt one position or another

It covers hybrids, EVs (briefly), flex fuel, displacement on demand, diesels (old news for Europeans but less common in the US), autogas and hydrogen. The book is an easy read: clearly written, nicely laid out and with lots of space on the page.

## Electrical Contacts: Principles and Applications (Electrical and Computer Engineering)

by: Slade, Paul G.          Date: 1999      ISBN: 824719344

Published by: CRC, Edition: 1, Hardcover, 1104 pages

This is a serious academic work of over 1000 pages and covers a range of topics in detail from the mechanical and electrical issues in connector design to arcing in high-current switches. There is a lot of detailed information on the behaviour of different kinds of materials and platings. The book introduced me to the idea of 'fretting' (the build up of insulating oxides in a connector resulting from wear under vibration) and discusses the benefits of using lubricants on connectors: although the discussion is fairly academic for the general user. There is a discussion of brush design relevant to DC motors with a reminder of the trade off between electrical efficiency and brush life.

# Limited usefulness

## Electric and hybrid vehicles : design fundamentals

by: Husain, Iqbal       Date: 2003      ISBN: 849314666

Published by: London : CRC Press

Despite its title, this book is emphatically not for someone trying to understand the basics of EVs and HEVs. It is highly mathematical, and the qualitative descriptions are thin. A good example is the explanation of Peukert's number on page 68/69. This is almost wholly mathematical with no hint of its significance (a high Peukert number is an issue in an EV battery because it means the battery won't perform as well as expected at the high discharge rates used).

The book might be helpful to someone who was familiar with the concepts, had excellent mathematical skills and planned to build an EV simulator. I stress "might" because I do not know enough to validate the accuracy of the mathematics. Even at a qualitative level I did not have a great deal of confidence in the book. The coverage of engine and motor sizing for an hybrids in the final chapter seemed to me to be less incisive than I have encountered elsewhere, and possibly mistaken

## Novel cathodic materials for rechargeable Lithium batteries

by: Koltypin, Maxim      Date: 2011      ISBN: 3639356993

Published by: VDM Verlag Dr. Muller, Paperback, 64 pages

This book has little useful content. It looks suspiciously like recycled material from the author's PhD thesis from 2006 which makes it an historical text in this field.

## Build Your Own Electric Vehicle

by: Leitman, Seth      Date: 2008      ISBN: 71543732

Published by: McGraw-Hill Professional, Edition: 2, Paperback, 327 pages

This is Leitman's update of Bob Brant's original 1993 book (see above). Like the original book it suffered from excessive tree-hugging. I read the book for an insight into electric vehicles, not a rehash of material from the International Panel on Climate Change. Given its date, I was surprised that there was not more discussion of Lithium batteries and AC drives.

Also, do not expect a lot of detail on the actual process of conversion. There was just one chapter on this and it was quite general - not clearly based on a single conversion exercise, and certainly nothing like detailed step-by-step instructions. Even the theory chapters were a bit shallow and lacking both clarity and authority. If you were expecting to come away with a clear understanding of the difference between a wound-rotor and a squirrel cage motor, or the pluses and minuses of different Lithium chemistries, you would be disappointed.

## Build Your Own Plug-In Hybrid Electric Vehicle (Tab Green Guru Guides)

by: Leitman, Seth      Date: 2009¬¬629.229   ISBN: 71614737

Published by: McGraw-Hill/TAB Electronics, Edition: 1, Paperback, 320 pages

This book is riddled with basic misunderstandings (e.g. lack of clarity about the distinction between torque and power). One graph for example purports to show the power output of series DC motors but uses a typical torque curve instead, showing power as a maximum at zero r.p.m. (a mathematical impossibility) This is like finding a book on climate confusing heat and temperature: it undermines any confidence in anything else the author says. The text was poorly structured both in terms of the flow of the topics and the depth at which they were tackled: you would read a page of general description then turn the page to find a circuit diagram from a Prius.

Finally I found the environmental finger-wagging irritating: I bought the book wanting to understand more about the engineering of hybrids and the environmental bits struck me as both inappropriate and partisan.

## The Electric Car: Development and Future of Battery, Hybrid and Fuel-Cell Cars (Iee Power & Energy Series, 38)

by: Westbrook, Michael H.          Date: 2001          `ISBN: 852960131

Published by: Institution Electrical Engineers, Edition: 1st, Hardcover, 224 pages

This is a review of battery electric vehicles, which also covers fuel cell and hybrid cars. By the time I read it (2008), it was already a little dated, with some predictions looking mildly quaint. The style is a bit dry and much of it requires at least A-level physics. Don't bother unless you are happy about the distinction between (say) Watts and Watt-hours

# Index